3D Print: Lighter Drones, More Payload

Shiva

Contents

Chapter 1

Introduction

1.1 Background to the Project

There have been many advances in additive manufacturing (AM) technologies over recent years. These technology advancements have enabled the creation of numerous lightweight antennas for a vast variety of applications that require very lightweight sensors. An example of this is in drone technology where a drone may only be capable of carrying a payload of a few hundred grams. In such a case, a conventional metallic antenna would in all likelihood be too heavy for the platform and render the application unrealisable.

The traditional antenna manufacturing methods often involve casting or precision milling. Such manufacturing processes are laborious and can also be very expensive when manufacturing the small quantities required for prototyping. For example, commercial horn antennas are expensive due to the material used and the use of precision machining. The dimensions of the antenna waveguides are critical to ensure the antenna satisfies their stringent requirements, and can be complicated to manufacture.

The development of AM technologies has seen the proliferation of creating radio-frequency (RF) and microwave components using these advanced techniques [1]. Fused deposition modelling (FDM) is one such 3D printing technology that has been utilised in the fabrication of microwave components. FDM-based 3D printers create parts using plastic and the plastic parts can be metallised to produce microwave components comparable to expensive commercial products.

1.2 Project Requirements

This book aims to investigate the feasibility of 3D printing antennas in a low-cost man-ner using a desktop 3D printer. A literature review on 3D printing technologies with relation to microwave component manufacturing applications is conducted. Basic horn antenna designs are required to be manufactured in order to investigate the process of 3D printing using plastic combined with a metallisation process. A full electrical characteristic analysis of the manufactured antennas is required and the results and conclusions thereof presented.

1.3 Research Objectives

The main objectives of this book are to:

- Conduct a literature review on antenna theory, additive manufacturing technologies, and 3D printing of antennas.

- Design pyramidal and conical horn antennas using a suitable simulation package, such as FEKO.

- Fabricate the designed antennas using 3D printing technology, and investigate the metallisation process.

- Measure and analyse the performance of the fabricated antennas.

- Produce a comprehensive report on the results of the fabrication process as well as the measurements.

These objectives are subject to strict budget limitations, as well as the availability of fabrication and testing facilities.

1.4 Software Tools

Various software tools have been used throughout this work and are outlined below:

- Altair FEKO 2017 - A computational electromagnetic software, developed by Altair Engineering, that is widely used in the telecommunications, automobile, space and defence industries. In this project FEKO is used to simulate all the antennas and provide their respective antenna properties. FEKO can also export the measurements into a csv file allowing other software tools to use the data, in this case Matlab.

- SolidWorks Premium 2015 - A 3D modelling computer-aided design (CAD) software. It is a solid modeller and uses feature-based parameterisations to create models and assemblies. It allows 3D models to be saved as STL files which can be used by 3D printer preparation software.

- Matlab v2018a - A numerical computing programming language developed by MathWorks. For this project Matlab is used to read csv files containing the raw data from both FEKO and the measured data which is then used to plot the results. Important properties can be extracted from the data and graphs, and the graphs also provide valuable and visual comparisons between different antenna measurements.

1.5 book Overview

Chapter 1 provides an introduction to the project and the motivation for the investigation of 3D printing antennas for the purpose of fabricating lightweight, low-cost antennas for rapid prototyping and experimental applications. The research objectives are stated and an overview of the book is presented.

Chapter 2 reviews the fundamental antenna theory required for the design of waveguide antennas, looking specifically at pyramidal and conical horn antennas. This chapter also discusses the current 3D printing technologies available and methods of metallising plastics, as well as conducting a comprehensive literature review of 3D printing microwave and RF components.

Chapter 3 discusses the design procedure where the specifications of the available 3D printer

Ultimaker 2+, and the choice of filament material used for 3D printing are discussed. The methods of replicating a commercial X-band horn and designing the Ku-band pyramidal and conical horns are described.

Chapter 4 describes the fabrication process of each of the designed horn antennas using the Ultimaker 2+. The metallisation procedure of the horns by electroless and electrolytic plating as well as the use of conductive painting is described. The methodology of measurement of the antennas' properties in the anechoic chamber is also described, and a cost analysis of the fabricated antennas is presented.

Chapter 5 presents all the measured antenna properties. The X-band pyramidal horn and Ku-band pyramidal and conical horns are presented with their measured reflection coefficients, isotropic gains, and co-polarised and cross-polarised radiation patterns. The chapter concludes by presenting and discussing the measured results.

Chapter 6 presents conclusions on the fabricated X-band and Ku-band antennas, as well as the implications of using the Ultimaker 2+ 3D printer and the different metallisation processes used. The chapter then concludes by summarising the costs of fabricating the antennas and provides recommendations for future work related to the project.

Chapter 2

Theory and Literature Review

2.1 Antenna Theory Review

This section will briefly describe the fundamentals of antenna theory. By definition, an antenna is "a device used to transform an RF signal travelling on a conductor, such as waveguides and transmission lines, into an electromagnetic (EM) wave in freespace. Antennas demonstrate reciprocity, which means that an antenna will maintain the same characteristics if it is transmitting or receiving" [2].

An antenna radiation pattern defines the power radiated by an antenna as a function of the direction away from the antenna, and is observed in the far field or Fraunhofer region [3]. Measuring the radiation pattern in the far field region reduces the phase variation across the antenna under test (AUT). There are two principal field regions: the near field or Fresnel region, and the far field or Fraunhofer region. The boundary between these two regions is generally defined as:

$$R_{ff} = \frac{2L^2}{\lambda_\circ} \quad [m] \tag{2-1}$$

where L is the maximum dimension of the antenna, and $\lambda_\circ$ is the wavelength corresponding to the antenna's frequency of operation. The field regions are shown in Figure 2-1. The wavelength is determined by the equation:

$$\lambda_\circ = \frac{c}{f_\circ} \quad [m] \tag{2-2}$$

where c is the speed of light, and f_o is the frequency of operation. For larger antennas, near-field techniques can also be used to measure the antenna radiation pattern.

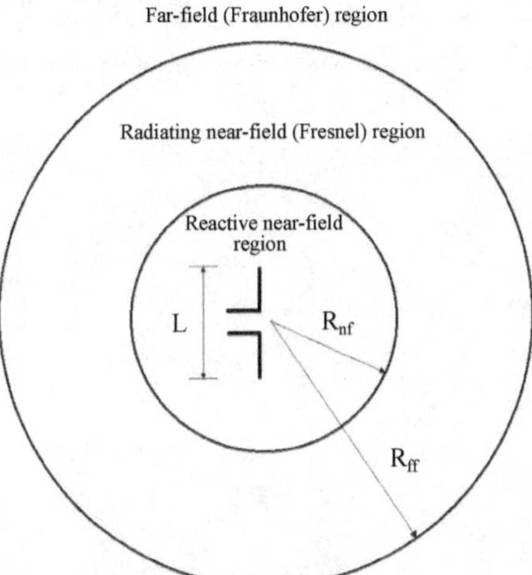

Figure 2-1: Field regions of an antenna [3].

Directivity describes how well an antenna can focus energy in a particular direction when transmitting, or how well it can receive energy from a particular direction when receiving [2]. It is a ratio of the maximum to the average radiation intensity, or, in the far field region, it may be expressed as the ratio of maximum to the average Poynting vector [4].

$$D = \frac{4\pi}{\Omega_A} \quad \text{[dimensionless]} \tag{2-3}$$

where D is the directivity and Ω_A is the beam solid angle.

Antenna efficiency is the ratio of the power delivered to the antenna relative to the power radiated from the antenna. The gain of an antenna is dependent on its directivity and antenna efficiency, and is generally given in reference to a lossless isotropic source:

$$G = \eta D \quad \text{[dimensionless]} \tag{2-4}$$

where η is the antenna efficiency.

Beamwidths and sidelobe levels are characterisations of the antenna radiation pattern. The beamwidth is a measure of the main beam at half power, or −3 dB from the peak power, also know as the half-power beamwidth (HPBW). Sidelobes represent radiation that is not in the direction of the main beam. The sidelobe level (SLL) of a pattern is measured as the difference between the peak of the main beam and the peak of the first sidelobe to either side of the main beam.

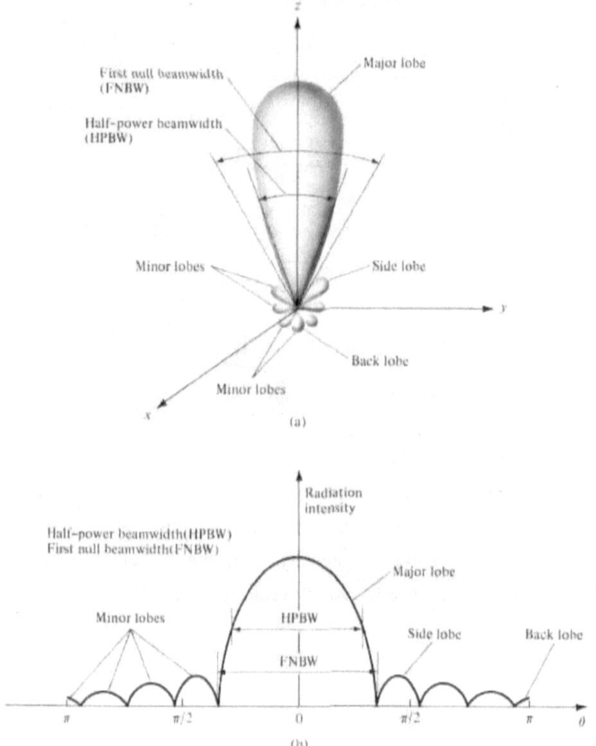

Figure 2-2: Radiation beamwidths and lobes of an antenna pattern [3].

The effective aperture of an antenna is a useful parameter to calculate the transmit and receive power of an antenna. It describes how much power is captured from a given plane wave, and factors in the intrinsic losses to the antenna. A general relation of effective aperture to antenna gain is given by:

$$A_e = \frac{\lambda_\circ^2}{4\pi} G \quad [m^2]$$

(2-5)

The Friis transmission equation is used to calculate the power received from one antenna, when transmitted from another antenna in a one-way propagation, based on the separation distance between the antennas and the operating frequency or wavelength.

$$P_R = \frac{P_T G_T G_R \lambda_\circ^2}{(4\pi R)^2} \ \ [\text{W}]$$

(2-6)

where

- P_R is the power received at the receiving antenna in Watts.

- P_T is the power transmitted by the transmitting antenna in Watts.

- G_T is the gain of the transmitting antenna.

- G_R is the gain of the receiving antenna.

- $\lambda_\circ$ is the freespace wavelength at the operating frequency in meters.

- R is the distance between the transmitting and a receiving antennas in meters.

In a one-way propagation, the power is dissipated at a factor of R^2.

2.1.1 S-Parameters

Scattering parameters or S-parameters are used to describe the input-output relationship between ports in an electrical system [5]. They are measured using network analysers and are represented using S_{mn} where 'm' denotes the input port and 'n' denotes the output port. S-parameters are usually measured in decibels, and can also be represented as a ratio between port m and n:

$$S_{mn} = 10^{\frac{S_{mn}[\text{dB}]}{10}}$$

(2-7)

An S-parameter value of less than $0\,\text{dB}$ indicates that there are power losses, whereas a value greater than $0\,\text{dB}$ indicates a gain. A commonly used S-parameter for antennas is S_{11} which represents the reflection coefficient that measures the amount of power reflected back to port 1 that has been transmitted from port 1. Another common measurement is the transmission coefficient

S_{21}, which represents the transmission efficiency between port 2 and port 1. The transmission coefficient can be incorporated into the Friis equation where the receive and transmit power ratio (P_R/P_T) is equivalent to S_{21}. This is particularly useful in solving gain.

2.1.2 Waveguides

Waveguides are used to transfer EM power efficiently from one point in space to another [6]. Coaxial cables, two-wire and microstrip transmission lines, and hollow conducting waveguides are common guiding structures, and are shown in Figure 2-3. The type of waveguide used is dependent on the desired operating frequency band, the amount of transferred power required, and the amount of tolerable transmission losses.

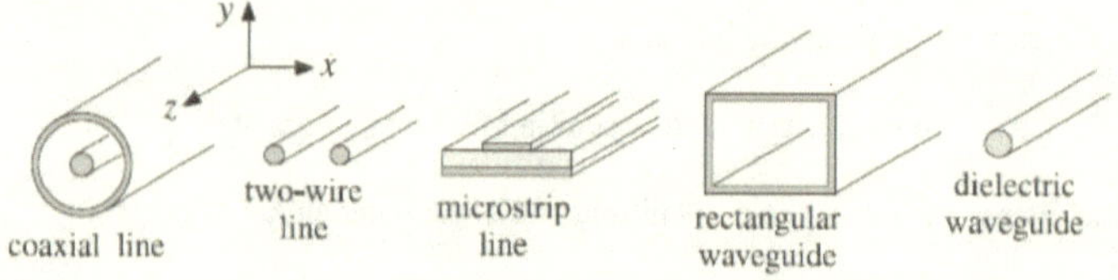

Figure 2-3: Typical waveguiding structures [6].

Waveguides can transfer EM waves in different formats. The mode of transmission corresponds to the different elements within an EM wave. The three common modes of transmission are:

- **TE$_{mn}$ mode:** The transverse electric mode is characterised by the electric (or E-field) component of the wave being always perpendicular to the direction of propagation.

- **TM$_{mn}$ mode:** In the transverse magnetic mode, the magnetic (or H-field) component of the wave is always perpendicular to the direction of propagation.

- **TEM mode:** In the transverse electromagnetic mode, both the E-field and H-field components are perpendicular to the direction of propagation.

The subscript 'm' represents the number of half wavelengths across the width (horizontal plane) and 'n' represents the number of half wavelengths across the height (vertical plane) of the waveguide aperture [7]. Waveguide propagation modes depend on the operating wavelength, polarisation, and the shape and size of the guide. These constraints are limited by the boundary conditions of wave propagation. The cutoff frequency of a waveguide is the lowest frequency at which the boundary conditions are still satisfied. The fundamental mode of the waveguide is the mode with the lowest cutoff frequency [6, 8].

This project only focuses on the fundamental modes of the waveguides. Figure 2-4 presents the cross section of the rectangular and circular waveguides when vertically polarised. The solid lines and dashed lines represent the E-field and H-field distribution respectively.

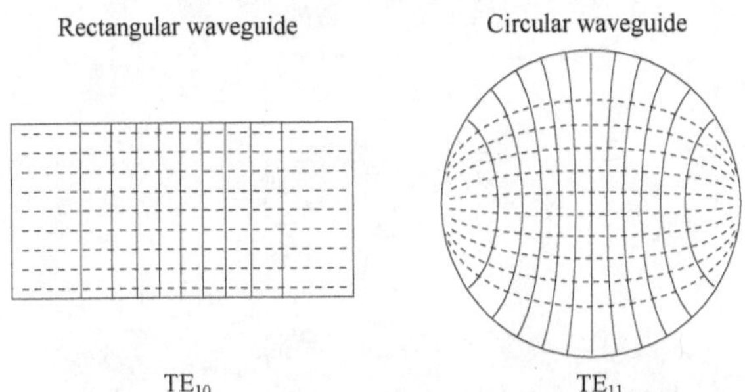

Figure 2-4: Fundamental modes for the rectangular (left) and circular (right) waveguides [9].

The fundamental mode for the rectangular waveguide is the TE$_{10}$ mode where the E-field distribution is one half wavelength across the broadside width and no wavelength across the narrowside width. The circular waveguide has a fundamental mode of TE$_{11}$ where the E-field distribution is one half wavelength across the vertical and horizontal planes. Note that in the TE$_{10}$ mode in the rectangular waveguide the E-field distribution is purely vertical, whereas in the TE$_{11}$ mode in the circular waveguide, the E-field distribution is curved. For both modes of propagation in their respective waveguides, the maximum E-field is at the centre of the waveguide, therefore most of the radiated power is concentrated at this maximum.

When waveguides are connected to antennas, it is important to consider their impedance matching, since this determines how well the wave is transmitted through the waveguide to the antenna. The impedance ratio is a measure of the impedance matching and determines how much of the propagated wave is transmitted and reflected. This is also known as the reflection coefficient and transmission coefficient that has been discussed in Section 2.1.1. In radio transmission it is essential to have good impedance matching from the waveguide to the antenna to achieve optimum power transmission and reception.

2.1.3 Polarisation

Polarisation of an EM wave refers to the directions of the E- and H-fields and is observed in the far field region. Figure 2-5 shows the E- and H-fields in linear polarisation where the direction of propagation is in the positive z-axis.

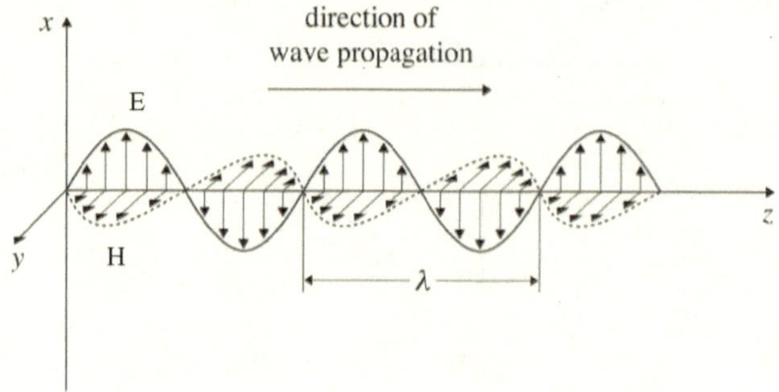

Figure 2-5: E- and H-fields shown in vertical polarisation [10].

The polarisation of a propagating EM wave corresponds to the orientation of the E-field of the wave. Linear polarisation is where the E-field oscillates back and forth on the same linear plane. If the E-field is parallel to the ground plane, it is known as horizontal polarisation, whereas if the E-field is perpendicular to the ground plane, it is known as vertical polarisation. Circular polarisation is where the E-field rotates in a circle, transverse to the direction of propagation. If the direction of propagation is out of the page and E-field rotates in a counter-clockwise direction, the field polarisation is right hand circularly polarised (RHCP), and if the E-field rotates in a clockwise direction, it is known as left hand circularly polarised (LHCP). Circular polarisation is a special case of elliptical polarisation where the magnitude of the E-field remains constant.

In order to maximise the received signal strength, the polarisation state of the receiving antenna should match that of the incoming wave, this is known as co-polarisation (co-pol). Cross polarisation (x-pol) is defined as the orthogonal pair with no reception where the E-fields of the transmitting and receiving antennas are orthogonal. Generally, x-pol is unwanted in transmission and reception of signals, but in reality an antenna is never fully polarised in one mode, thus the co-pol and x-pol radiation pattern are measured to observe the respective peak power and power leakage into the orthogonal polarisation [10, 11].

2.1.4 Pyramidal Horn

The pyramidal horn is one of the most commonly used antennas in radio frequency applications. Pyramidal horns are widely used because they can achieve high gain with a moderate bandwidth at the frequency for which they are designed to operate. They are also relatively simple to construct and they are often used as feeders for reflectors. Figure 2-6 shows the basic structure of a pyramidal horn.

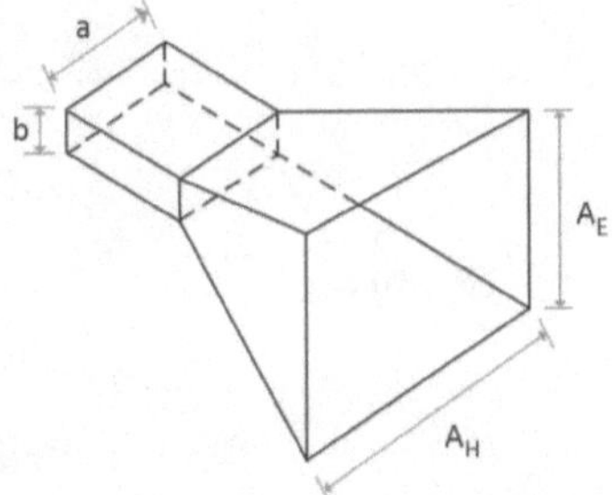

Figure 2-6: Structure of a pyramidal horn [12].

The basic structure of the pyramidal horn in Figure 2-6 has the following dimensions:

- a is the waveguide width

- b is the waveguide height

- A_E is the aperture width in the E-field direction

- A_H is the aperture width in the H-field direction

The pyramidal horn design has various parameters which are considered. Figure 2-7 shows some of these design parameters considered for the aperture of the horn. The aperture of the horn plays an important role in determining the HPBW and gain of the antenna. The gain of the antenna will vary with the slant angle which determines the aperture length.

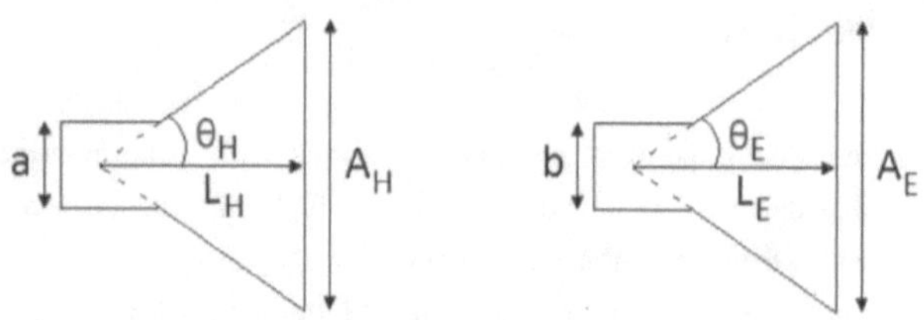

Figure 2-7: Cross-sectional view of a pyramidal horn showing aperture design parameters [13].

The theoretical design for the aperture lengths of the horn is governed by the following equations [14, 15]:

$$A_E = \sqrt{2\lambda_\circ L_E} \tag{2-8}$$

$$A_H = \sqrt{3\lambda_\circ L_H} \tag{2-9}$$

where

- L_E is the length from the apex to the centre of the aperture in the E-field direction

- L_H is the length from the apex to the centre of the aperture in the H-field direction

- θ_E is the slant angle in the E-field direction

- θ_H is the slant angle in the H-field direction

Equations (2-8) and (2-9) are the optimum formulas to design a flare that maximises the power radiated in the desired direction and the achievable gain. The flare of the horn antenna provides a smooth match between the waveguide and freespace.

A pyramidal horn antenna is typically fed using a coaxial to waveguide transition. Coaxial cables only support the TEM modes because it has a centre and an outer conductor. Rectangular waveguides support TE and TM modes of propagation, but the fundamental mode of propagation is the TE_{10} mode. The broadside width of the waveguide is equal to a half wavelength and there is no wave across the height. Ideally, only the fundamental mode of propagation should exist in the waveguide, and to achieve this the centre operating frequency must not be above the next higher order mode of propagation. Using the knowledge of the propagation modes in rectangular waveguides, dimensions of rectangular waveguides have been standardised for operation at different frequencies. Figure 2-8 shows the structure and parameters of a typical hollow waveguide.

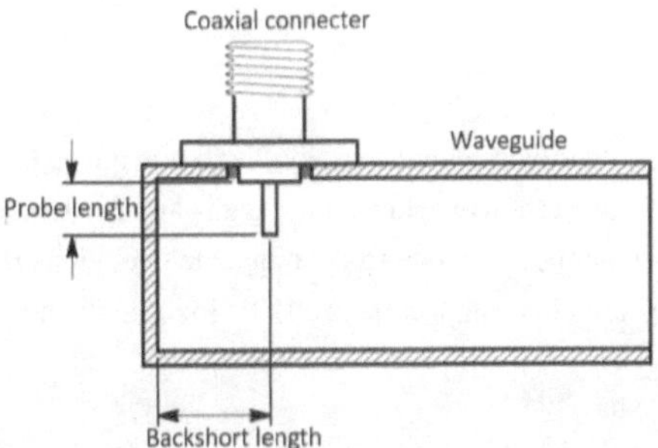

Figure 2-8: Typical waveguide structure and parameters [16].

The waveguide of the horn is designed such that the transition of the transmitted wave from the coaxial connector to the flare is well-matched, achieving good transmission efficiency. The waveguide wavelength, λ_g, is determined by [17]:

$$\lambda_g = \frac{\lambda_o}{\sqrt{1 - \left(\frac{\lambda_o}{\lambda_c}\right)^2}} \tag{2-10}$$

The cutoff frequency of the fundamental mode for a rectangular waveguide is given by:

$$\lambda_c = 2a \tag{2-11}$$

where a is the broadside width of the waveguide. Theoretically, a well-matched horn will have a waveguide feed structure that satisfies the following equations [18]:

$$L_g \geq \frac{3}{4}\lambda_g \tag{2-12}$$

$$l_p = \frac{\lambda_o}{4} \tag{2-13}$$

$$l_b = \frac{\lambda_g}{4} \tag{2-14}$$

where L_g is the waveguide length, l_p is the probe length, and l_b is the backshort length.

Equations (2-12), (2-13) and (2-14) are theoretical approximations and are generally used as initial parameters in designing the feed for a hollow waveguide. This ensures that the position of the probe creates constructive interference of the EM waves in the transition from the coaxial line to the waveguide, and that the EM wave propagates efficiently outwards of the antenna into freespace.

2.1.5 Conical Horn

Conical horns are also commonly used antennas, although it is not as frequently seen as the pyramidal horn. The conical horn can also achieve high gains and moderate bandwidths. They have been widely used in space applications due to their capable operations from the Megahertz to the Terahertz range, and they are also compact in size [19]. Figure 2-9 shows the basic structure of a conical horn.

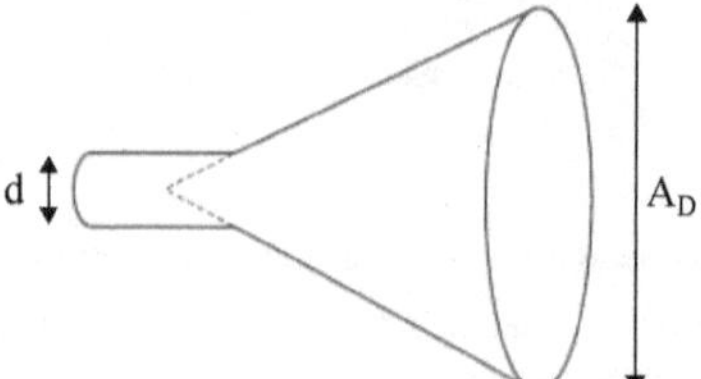

Figure 2-9: Structure of a conical horn.

where d is the waveguide diameter and A_D is the aperture diameter. Conical horns have a very simple structure, where a circular waveguide is usually fed by a coaxial line and the waveguide flares outward in a conical shape. The dimensions of the conical aperture are important design parameters for the horn. Figure 2-10 shows the parameters needed to determine the aperture of the horn.

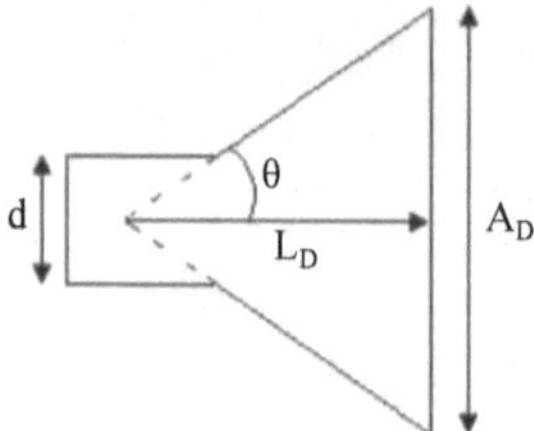

Figure 2-10: Cross-sectional view of the conical horn showing aperture parameters [13].

where L_D is the length from the apex to the centre of the aperture and θ is the slant angle.

In order for an EM wave to propagate most efficiently from the waveguide through the flare into freespace, the following theoretical optimum equation is used to determine the aperture dimensions [14, 15]:

$$A_D = \sqrt{3\lambda_\circ L_D}$$

(2-15)

where

- A_D is the aperture diameter

- $\lambda_\circ$ is the wavelength at operating frequency

- L_D is the length from the apex to the centre of the aperture

Conical horns are generally fed from a coaxial line, and use a circular hollow waveguide transition from coaxial to freespace. Similarly to the rectangular hollow waveguide, the feed structure of the circular waveguide is governed by optimum formulas that are used as a base guideline for design purposes. The waveguide wavelength can be calculated using the same equation in (2-10):

$$\lambda_g = \frac{\lambda_\circ}{\sqrt{1 - \left(\frac{\lambda_\circ}{\lambda_c}\right)^2}}$$

The wavelength of the lower cutoff frequency, which is also the fundamental mode cutoff frequency for a circular waveguide, is given as [20]:

$$\lambda_{c(low)} = 1.706d$$

(2-16)

While the wavelength of the upper cutoff frequency is given as:

$$\lambda_{c(high)} = 1.306d$$

(2-17)

where d is the waveguide diameter. The cutoff frequency of a waveguide is very important, and it sets a limit on the lowest frequency at which the fundamental mode of the EM wave can propagate through the waveguide. In accordance with any waveguide, it is ensured that the EM wave propagates most efficiently through the waveguide, and that no higher order modes are introduced. This is done by designing the waveguide such that the centre operating frequency is between the upper and lower cutoff frequency. The fundamental mode for a circular waveguide is the TE_{11} mode where there is a half wavelength in both the width and height of the waveguide aperture.

2.2   Skin Depth

When EM waves propagate through a medium, the amplitude of the wave will attenuate as a function of the properties of the medium. Skin depth, or depth of penetration, is the depth at which a wave has been attenuated to 1/e, i.e., approximately 37% of its original value [21, 22]. This is a measure of how deep a wave penetrates into a material. Skin depth, σ, can be determined using the following equation:

$$\sigma = \sqrt{\frac{\rho}{\pi f_\circ \mu}} = \sqrt{\frac{\rho}{\pi f_\circ \mu_r \mu_0}} \qquad (2\text{-}18)$$

where

- $f_\circ$ is the operating frequency of the EM wave

- ρ is the resistivity of the material

- μ is the permeability of the material

- μ_0 is the permeability constant

- μ_r is the relative permeability

When a wave is incident on the surface of a material, part of the wave may be reflected and another part is transmitted into the material. The conductivity and permeability of the material, as well as the wavelength of the incident wave, determine how far into the material the wave travels. This means for the same material, skin depth will be small at high frequencies, and large at low frequencies. In a horn antenna for example, the thickness of the conductive surfaces through which the EM wave propagates, affects how well the wave "bounces" off the surfaces. The frequency of the propagating wave also contributes to the skin depth. A general good rule of thumb for conductive surfaces is to have a minimum thickness of six times the skin depth to ensure that the propagating EM wave does not penetrate the entire thickness of the material [23].

2.3   3D Printing

In recent years, 3D printing has become more prominent in RF applications and very useful for prototyping. The advancement in 3D printing technologies has created faster and cheaper methods of fabricating parts, as well as parts with complex designs which previously would have been more difficult to fabricate through traditional means, such as CNC machining. 3D printing or additive manufacturing, also known as rapid prototyping, is defined by the American Society of Testing and Materials (ASTM) as "the process of joining materials to make objects from 3D model data, usually layer upon layer, as opposed to subtractive manufacturing methodologies." The ASTM classifies AM technologies into several categories [24].

Process categories	Technology	Materials
Binder Jetting	3D Printing Ink-jetting S-Print M-Print	Metal Polymer Ceramic
Direct Energy Deposition	Direct Metal Deposition Laser Deposition Laser Consolidation Electron Beam Direct Melting	Metal: powder and wire
Material extrusion	Fused Deposition Modeling	Polymer
Material Jetting	Polyject Ink-jetting Thermojet	Photopolymer Wax
Powder bed fusion	Selective Laser Sintering Selective Laser Melting Electron Beam Melting	Metal Polymer Ceramic
Sheet lamination	Ultrasonic Consolidation Laminated Object Manufacture	Hybrids Metallic Ceramic
Vat photopolymerization	Stereolithography Digital Light Processing	Photopolymer Ceramic

Figure 2-11: Classification of additive manufacturing technologies [24].

The categories in Figure 2-11 are binder jetting, direct energy deposition, sheet laminations, material extrusion, powder bed fusion, and vat photopolymerisation. Each category has its own distinct process but they all share the same principle of layer selective modelling. There is a wide variety of materials that can be used for AM, and is dependent on the AM process that is used.

Currently, material extrusion, material jetting, vat photopolymerisation, and powder bed fusion are the most investigated AM techniques for satellite applications [24].

2.3.1 Material Extrusion

Material extrusion uses the technology of fused deposition modelling (FDM), which entails the extrusion and subsequent deposition of a molten filament of polymeric material [24]. Figure 2-12 shows the scheme of the FDM process.

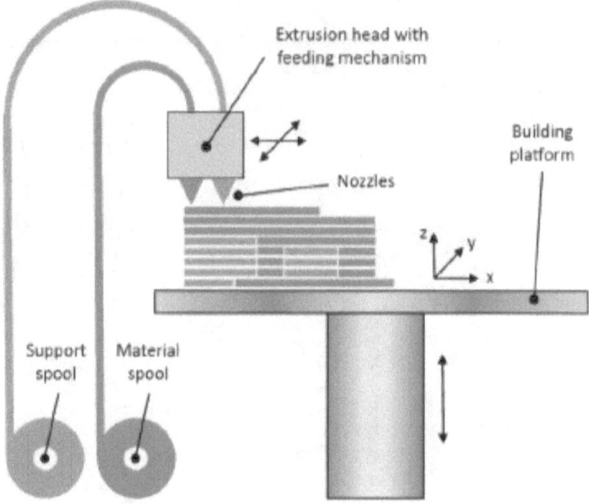

Figure 2-12: Scheme of the fused deposition modelling process [24].

The physical principle of the FDM process is very similar to that of a hot glue gun. A filament, usually of polymeric material, is softened and melted by heat and pushed through a nozzle and deposited layer by layer onto a building platform or bed.

In comparison to other traditional manufacturing processes, 3D printing technology has far less of an impact on the environment. 3D printers generate less waste by using a little more than the amount of material necessary for the product, completely eliminating the process of drilling, cutting, and milling. They demonstrate considerable advantage over traditional machines in terms of carbon footprint.

2.3.2 Material Jetting

Material jetting is a process that creates 3D models by moving inkjet print heads that jet photopolymer onto a building platform [24]. Figure 2-13 shows this technology where one or more print heads move over the build area, and droplets are selectively ejected onto the build platform.

This selective jetting technique allows positional high accuracy, low waste, and small droplet size which allows wider material availability. Ultraviolet (UV) light is used to cure the deposited material and the model is created layer by layer. Jetting requires materials to be heated, thus it is common for solvents to be used in the fluid.

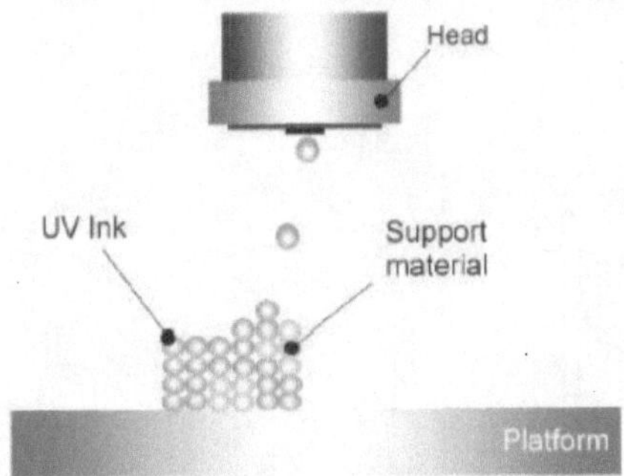

Figure 2-13: Scheme of the material jetting process [24].

Jetting also facilitates deposition of different colours and hardness of materials onto the same part, which can create components with varying properties and characteristics. Support structures are used in jetting and are often built in a different material to the component. The support material can be removed using a water jet or a sodium hydroxide solution. The high accuracy of the material jetting process allows for minimal post-processing.

Material or inkjet printing is most commonly still used in commercial applications such as graphics, product marking and coding, although in recent years the technology is becoming more prominent in technological areas such as aerospace and defence, automotive, architecture, healthcare, sport and entertainment [24].

2.3.3 Vat Photopolymerisation

Vat photopolymerisation is defined by the ASTM subcommittee in 2012 as "an additive manufacturing process in which liquid photopolymer in a vat is selectively cured by light-activated polymerisation". Stereolithography (SL) is the most common vat photopolymerisation process. The principle of the SL process is the polymerisation of a photosensitive resin. Figure 2-14 demonstrates the scheme of a vat photopolymerisation process, or the stereolithography process.

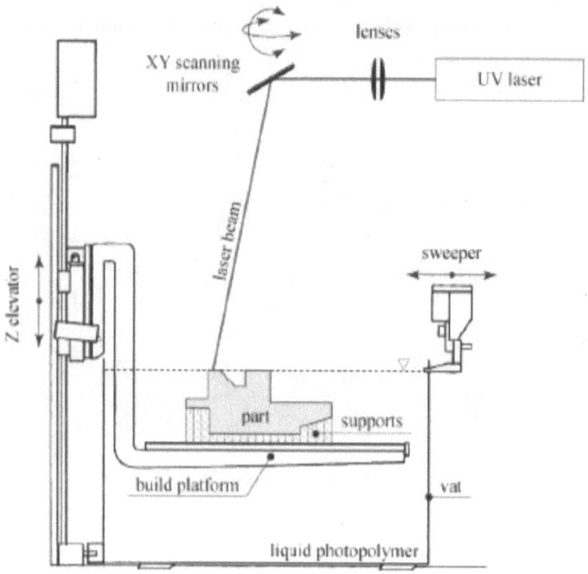

Figure 2-14: Scheme of the vat photopolymerisation process [24].

A building platform is lowered into a resin bath by a layer thickness, and an ultraviolet laser selectively cures the resin, converting it to a layer of solid material. The platform is then lowered into the resin again, and a new layer is built on top of the previous one. Support structures are made from the same material, as this SL scheme can only process one material. The same process is applied in different machines where the part is built from the underside of the build platform in a top-down layer-by-layer manner.

SL commonly requires post processing of the part, where the part will be cleaned to remove excess liquid resin and to detach supports. The part will then complete its curing process in a UV oven. Generally in SL, the layer boundaries of the part are first created and the internal areas are created in hatching patterns. Exposure of the photopolymer to UV light causes the photopolymer to shrink, and this results in the part to curl on its edges which can cause stresses to remain in the part. To avoid this, a star-weave pattern can be used in conjunction with multiple scans of each layer [25].

The SL process allows fabrication of parts with very smooth surfaces and a high level of accuracy, and the process is relatively quick. However, support structures are often required, which makes post-processing operations longer and more labour-intensive. SL can be used with a wide range of materials which can accommodate many different applications. SL parts can also be metallised to enhance the strength while still keeping the part lightweight.

2.3.4 Selective Laser Melting

Selective laser melting (SLM) is a process which involves a powder bed fusion using a laser to melt the powder layer by layer to build a part from the bottom upwards [24]. The process of an SLM machine is shown in Figure 2-15.

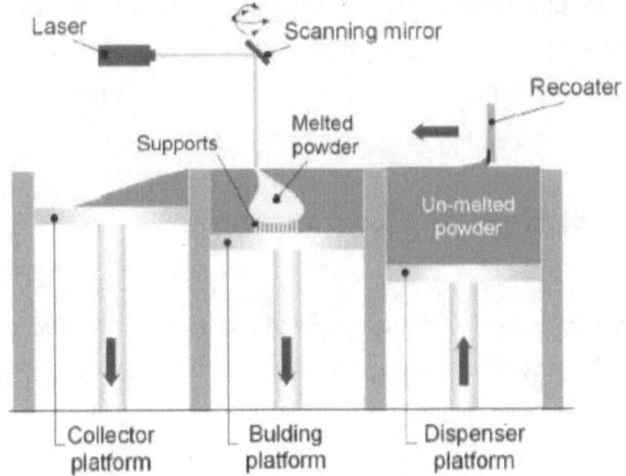

Figure 2-15: Selective laser melting machine process [24].

Like most other additive manufacturing methods, in SLM a 3D model is processed by software to determine the need for and provide support structures, as well as generate slice 2D cross-section data for the laser scanning of individual layers. Metal powder is laid in a building chamber on a substrate build plate, and a laser selectively fuses areas according to the provided data. The platform is then lowered and the laying of powder and laser scanning repeated for each new layer until the part is complete.

The SLM process can be optimised by adjusting process parameters such as the laser power, scanning speed, hatching distance, and layer thickness. Various combinations of these parameters can affect the mechanical properties and surface roughness of the produced parts. While the effect can improve some of the properties, the resulting surface quality is often still inferior to that of a conventionally manufactured part. Various post-processing techniques are employed to improve surface quality, but it is not applicable to more complex parts.

There are currently many alloys available for the SLM process, including stainless steel, cobalt chromium, aluminium, and titanium. Stainless steel is the most popular alloy used in SLM due to its versatile applications in biomedical equipment, marine, and fuel cells [26, 27, 28, 29].

2.4 Metallising Plastics

The process in which plastics are coated with metal, is called metallisation. Applications of this process are generally for mechanical or aesthetic purposes, but this metallisation process can also be used to turn plastic models into functional devices, such as antennas and filters. Metallisation of plastic can be done using several methods, such as plating, or painting.

2.4.1 Electroless Plating and Electroplating

Metallising plastic can be achieved by using the electroless plating and electroplating method. Electroless plating is the process of plating parts without using an external power source - it is a chemical process.

ABS is the most widely used plastic in the plating industry. It is most suitable for plating because there is no mechanical abrasion required to deposit adherent metal coating on it - only the use of chemical pretreatment is required before electroplating [30].

Plating of ABS generally follows the process as depicted in Figure 2-16 which can be divided into two major steps: surface treatment or preparation, and metal plating. The surface preparation stage involves etching, neutralisation, activation and acceleration, which are chemical processes. Etching roughens the surface and enhances the surface area, by creating micro-pores which serve as bonding sites between the substrate and the metal. The neutralisation process removes any excess etchant (chemical used for etching). The activation process produces the catalytic surface required for plating by depositing an activating agent such as palladium onto the surface. The acceleration process removes excess chemicals surrounding the activating agent (palladium) and keeps the surface intact for electroless plating.

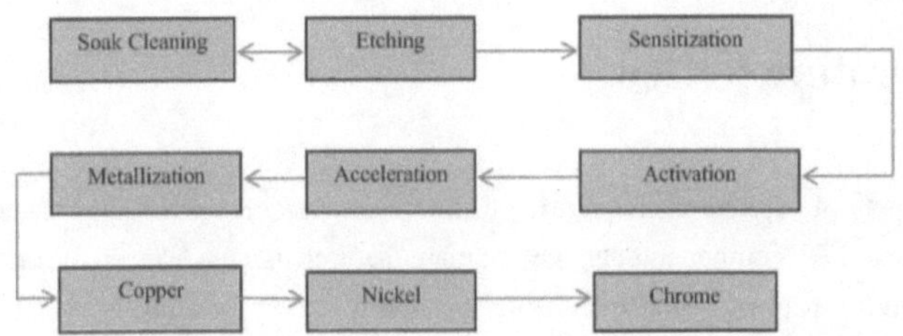

Figure 2-16: Schematic diagram representing the electroplating process of ABS [30].

The metal plating stage involves metallisation, copper coating, nickel coating, and finishing. The metallisation process is where electroless plating occurs. Copper is deposited onto the surface, and continuous metal layers are created. Thereafter, electrolytic plating or electroplating can be carried out to increase copper layer thickness and give a better conductive surface. Electrolytic nickel plating is commonly carried out after the copper plating stage to prevent the corrosion of the copper coating. In the finishing stage, the coated part can be finished with gold, silver, brass, chrome, etc., as required.

2.4.2 Conductive Painting

Metallising plastics can also be done by applying conductive paint onto the plastic. Applying conductive paint onto plastic is widely used in the electronics industry for electromagnetic interference (EMI) or radio frequency interference (RFI) shielding as well as electrostatic discharge (ESD) prevention.

Conductive paint can be applied onto the plastic using the traditional painting method by brushing the paint onto the surface, but it is more common now to use atomisers or spray equipment. In the manufacturing industry, paint is commonly applied using spray guns. Spray painting is more advantageous than brush painting because it is faster to apply the paint, and the paint will dry faster. Spray painting also uses less paint, thus the cost of material used will be cheaper.

Copper, silver, nickel and graphite conductive paints are commonly used in EMI/RFI shielding applications. The application of these conductive paints can be performed by entities using commercial-grade equipment. Such conductive paints are also commercially available in the form of aerosols.

2.5 Literature Review

The advancement of additive manufacturing technologies in recent years has created new avenues in the RF industry. RF components can now be manufactured using advanced 3D printing methods, either by printing in plastic and metallising the plastic, or by printing in metal. This chapter will review selected literature on fabricating RF components such as antennas and filters that use additive manufacturing technologies.

2.5.1 Ka-Band 3D Printed Horn Antennas

In [31], Yao et. al. presents 3D printed pyramidal horn antennas operating at 28 GHz. The antennas are based on a WR-28 waveguide standard gain horn designed for 20 dBi gain. The antennas have been printed in ABS material using an FDM printing system, and metallised using two methods: conductive copper paint, and copper tape, both applied to the 3D printed surface.

Figure 2-17 shows a photograph of the fabricated horn antennas for operation at 28 GHz. The measured reflection coefficient of the horn antennas have been obtained, and the measurements indicate that the antenna coated with copper conductive paint has a better reflection coefficient performance. The radiation patterns of the horns have been achieved in simulation and indicate satisfactory performance at 28 GHz.

Figure 2-17: Photograph of fabricated 20 dBi horn antennas operating at 28 GHz: left horn with copper conductive paint and right horn with copper tape on the surface [31].

An X-band horn antenna operating at 9.5 GHz has also been fabricated using 3D printing technology and coated in copper conductive paint. The gain has been measured, and the radiation pattern is comparable to the reference.

2.5.2 Laser-Based Layer-by-Layer Polymer Stereolithography for High-Frequency Applications

In [32], Maas et. al. uses the additive manufacturing method of laser-based layer-by-layer polymer stereolithography (SL) to create 3D structures for high-frequency applications. Polymer SL is shown to demonstrate favourable fabrication tolerances, which is ideal in making miniature RF components and narrowband filters. Using this SL technology, prototype components such as resonators, filters and antennas have been made.

Figure 2-18 shows a 2 GHz to 12 GHz double-ridged horn antenna manufactured using SL, and Figure 2-19 shows the same antenna that has been coated with a layer of conductive silver ink. The measured return loss of the antenna within 2 GHz to 12 GHz is better than 9.2 dB. The simulated and measured gains indicate the antenna efficiencies across 2 GHz to 12 GHz are greater than 87 %. The radiation patterns of the antenna has been measured at several different frequencies, and the measured results agree well with simulations.

Figure 2-18: 2 GHz to 12 GHz double-ridge horn antenna fabricated using stereolithography [32].

Figure 2-19: 2 GHz to 12 GHz double-ridge horn antenna metallised using conductive silver ink [32].

Other demonstrations of SL in microwave components and antennas include high-Q cavity resonators at K-band, low-loss narrowband microwave filters at K-band, an integrated cavity resonator with monopole at K-band, and evanescent-mode filters at X-band. Ion trap miniaturisation has also been demonstrated using SLA.

Maas et. al. concludes that the fabrication tolerances of SLA are excellent for fabricating RF devices operating up to K-band, whereas the fabrication of iron traps of the same tolerances exhibit larger effects due to the smaller size of the iron trap.

2.5.3   Selective Laser Melting Manufacturing of Microwave Waveguide Devices

In [33], Peverini et. al. investigates SLM for the fabrication of high-performance microwave waveguide devices. Using the SLM technology, fifth-order Ku/K-band filters and sixth-order Ku/K-band filters have been manufactured. The predicted and measured parameters of the filters have been compared and show good agreement. The weight of the fifth-order Ku/K-band filter manufactured through SLM is 27 g, whereas the same filter manufactured through conventional machining techniques increases the weight of the filter to 70 g. Figure 2-20 and Figure 2-21 respectively show the fifth- and sixth-order Ku/K-band filters fabricated using SLM technology.

Figure 2-20: Fifth-order Ku/K-band filter fabricated using the SLM printing method [33].

Figure 2-21: Sixth-order Ku/K-band filter fabricated using the SLM printing method [33].

Peverini et. al. concludes that SLM is a promising method for the manufacturing of all-metal waveguide components, and indicates further research will involve developing mechanical- and electromagnetic-coupled designs to implement waveguide components with enhanced accuracy in aluminium and titanium.

2.5.4   Application of Direct Metal Laser Sintering to Waveguide-Based Passive Microwave Components, Antennas, and Antenna Arrays

In [34], Chio et. al. investigates the application of direct metal laser sintering (DMLS) to waveguide-based passive microwave components, antennas, and antenna arrays. Designs for weight reduction have been investigated, as well as designs for complicated and intricate internal structures.

Figure 2-22 shows an X-band antenna fabricated using DMLS technology for the purpose of weight reduction. Figure 2-23 shows a lightweight, wideband hybrid waveguide coupler, and Figure 2-24 shows a novel patch array comparing effects due to perforation.

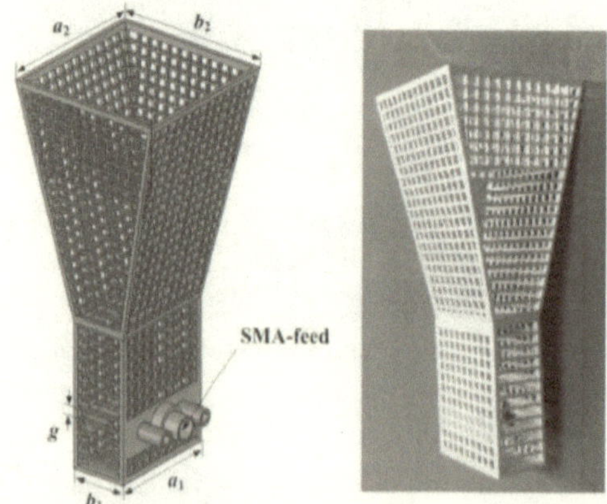

Figure 2-22: A lightweight X-band horn fabricated via direct metal laser sintering [34].

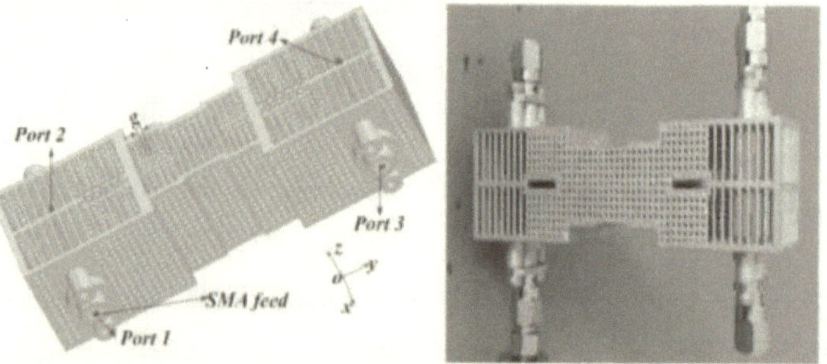

Figure 2-23: A lightweight, wideband hybrid waveguide coupler fabricated via direct metal laser sintering: (left) design; (right) prototype [34].

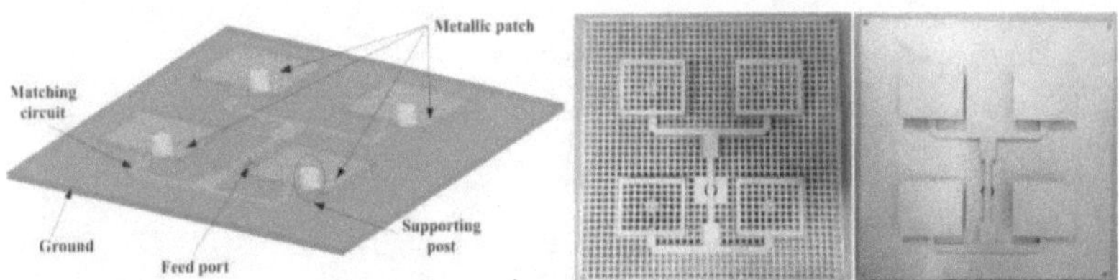

Figure 2-24: Novel patch array designed with DMLS in mind, comparing effects due to perforation: (left) design; (right) prototypes [34].

Figure 2-25 shows a corrugated horn with integrated septum polariser printed using DMLS. The horn has been designed for operation between 9.4 GHz and 9.8GHz, but performed well over a much wider bandwidth.

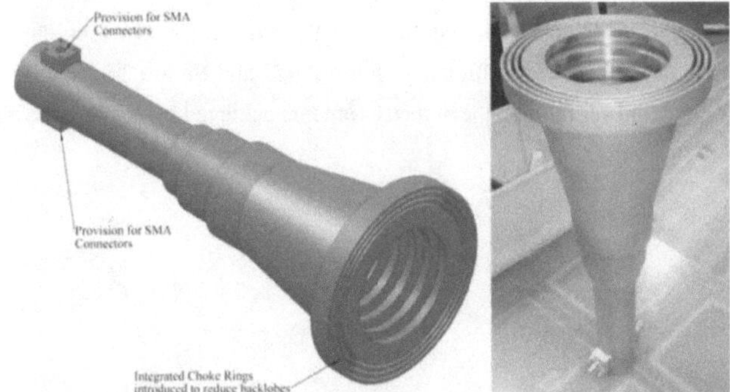

Figure 2-25: DMLS-printed dual-CP corrugated horn with septum polariser: (a) 3D drawing; (b) prototype [34].

Chio et. al. concludes that DMLS shows great versatility in its capabilities of making waveguide-based components, antennas, and antenna arrays. DMLS offers simple fabrication, faster turnaround times, and reasonable manufacturing costs.

2.5.5 Hyperband Bi-Conical Antenna Design Using 3D Printing Technique

In [35], Andriambeloson and Wiid manufactured an antenna using a 3D printing technique and conductive paint. A bi-cone antenna has been manufactured using this method and its electrical characteristics are evaluated. Figure 2-26 shows the 3D printer, the finished bi-cone antenna and the antenna feed description.

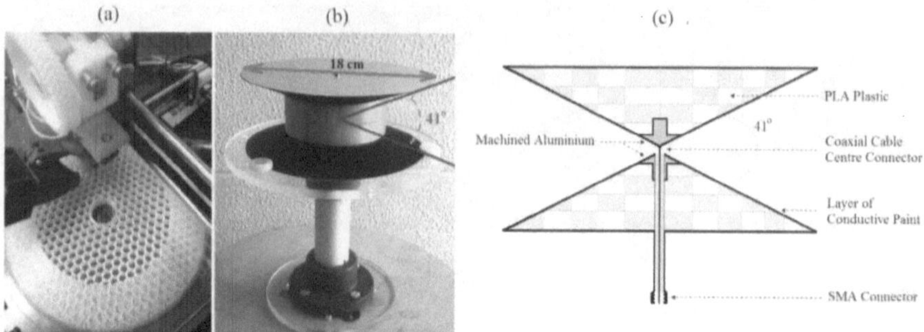

Figure 2-26: Overview of the 3D printed bi-cone antenna manufacturing: (a) shows the 3D printer, (b) displays the finished bi-cone antenna and in (c) the antenna feed is described [35].

The antenna parameters that have been measured are the reflection coefficient, the electric far-field directivity pattern, and the antenna efficiency. Figure 2-27 and Figure 2-28 respectively show the reflection coefficient and radiation pattern measurements achieved by the bi-cone antenna.

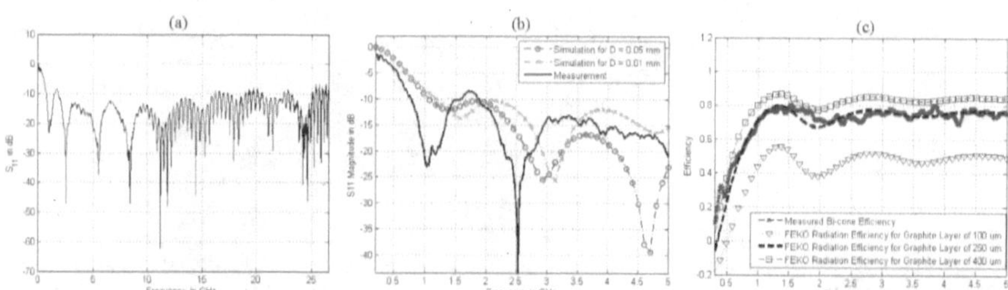

Figure 2-27: Reflection coefficient of the bi-cone antenna: (a) shows the measured S11 using a 26.5 GHz Agilent PNA-X VNA and in (b) it is compared with simulated curves up to 5 GHz. (c) displays the measured and the simulated antenna radiation efficiencies [35].

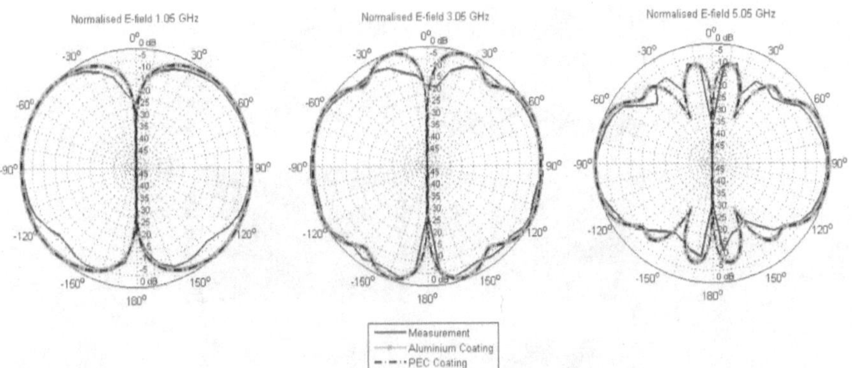

Figure 2-28: Comparison between measured and simulated bi-cone normalised EFD for 1.05 GHz, 3.05 GHz and 5.05 GHz [35].

Andriambeloson and Wiid [35] show that the 3D printing technique and the conductive paint used for the bi-cone antenna design is effective, but further work is required to address the antenna unbalanced feed. It is suggested that investigation in a combination of L-plates and absorbers could attenuate the flow of common-current across the feed-cable external conductor.

2.5.6 Investigation on 3D Printing Technologies for Millimeter-Wave and Terahertz Applications

In [36], Zhang et. al. investigates 3D printing technologies for millimeter-wave (mmWave) and terahertz (THz) applications. Dielectric 3D printed passive mmWave and THz devices have been reviewed, and horn antennas and waveguides have been fabricated using 3D printing technology.

Figure 2-29 shows a comparison of six V-band horns fabricated using different methods and Figure 2-30 shows the corresponding inner surface roughness of the horns. By analysing the surface roughness, the MMP-treated CU-15Sn has been determined to be the optimum choice for further antenna fabrication.

Figure 2-29: 3D printed V-band horns: (a) binder jetted 316L stainless steel without treatment; (b) gold-electroplated binder jetted 316L stainless steel; (c) MMP-treated binder jetted 316L stainless steel; (d) manually polished SLM Cu-15Sn; (e) gold-electroplated SLM Cu-15Sn; and (f) MMP-treated SLM Cu-15Sn [36].

Figure 2-30: Roughness of 3D printed V-band horns' inner surfaces: (a) binder jetted 316L stainless steel without treatment; (b) gold electroplated binder jetted 316L stainless steel; (c) MMP-treated binder jetted 316L stainless steel; (d) manually polished SLM Cu-15Sn; (e) gold-electroplated SLM Cu-15Sn; and (f) MMP-treated SLM Cu-15Sn [36].

Figure 2-31 shows the E-, D-, and H-band horns fabricated using the optimum method to determine the frequency limit of metallic 3D printing. Figure 2-32 shows the D-band waveguides fabricated using the same method.

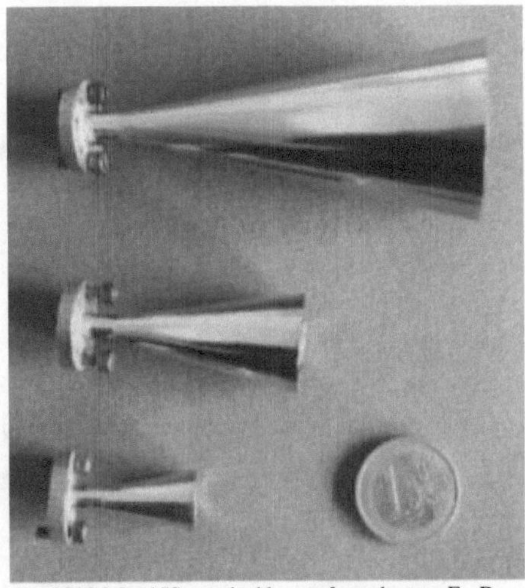

Figure 2-31: Photograph of SLM Cu-15Sn conical horns, from the top: E-, D-, and H-band horns [36].

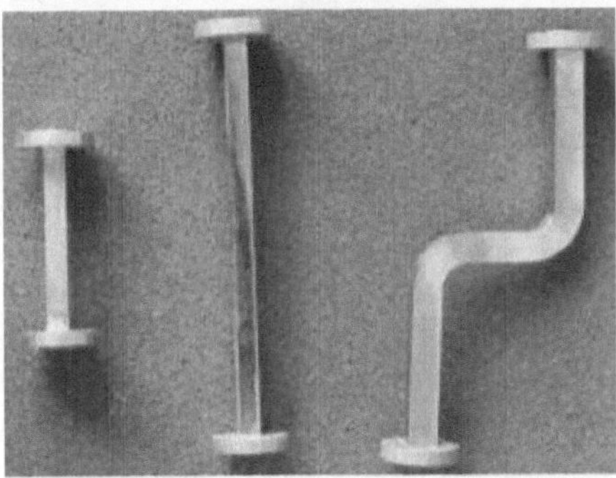

Figure 2-32: Photograph of SLM Cu-15Sn D-band waveguides, from left: 50 mm, 100 mm, and waveguide bend [36].

Zhang et. al. demonstrated that metallic 3D printing of mmWave and THz antennas can achieve comparable performance to traditional techniques such as micromachining, electrical discharge machining, and injection moulding. However, for the fabrication of waveguides, the 3D printing technology falls short when the waveguide size decreases and the structure complexity increases.

The most limiting factors of 3D printed devices in the mmWave and THz application are the surface roughness and dimensional tolerance.

2.5.7 Summary of Reviewed Literature

The fabrication of RF components and devices by means of 3D printing is still a relatively new area of research. Substantial work has already been done on metallising 3D printed antennas, but there is still much to be improved on and investigated. Based on the reviewed literature, it is noted that manufacturing antennas can be costly and labour intensive. Using the method of 3D printing can be more advantageous because the resulting part will have a lighter weight, lower cost, and also more bio-friendly. It is also a fast method of prototyping without having to employ costly manufacturing methods which can also be time-consuming. Thus, it is worth while to study the feasibility of fabricating antennas in a low-cost manner for the purpose of prototyping.

Chapter 3

Design Procedure

The aim of the project is to test the validity of a low-cost fabrication process of an antenna. The first step taken is to replicate a commercially manufactured antenna. An antenna has been chosen based on the printing capability of the available 3D printer. Thereafter, using the build volume of the 3D printer as a starting point, a frequency range of operation is chosen. With the premise of testing the process of fabricating an antenna using this method, a basic horn antenna in the Ku-band has been chosen. Other characteristics, such as the HPBW, gain, and bandwidth of the antenna is not specified, since the aim has been to test the fabrication process.

In aid of designing the antennas, the software package Altair FEKO 2017 has been chosen for antenna simulations. FEKO is a computational electromagnetics (CEM) software that is widely used in the telecommunications, aerospace and defence industries. FEKO hybridises several frequency and time domain EM solvers which enables efficient analysis of EM problems such as antennas and RF components.

In the following sections, the 3D printer used will be discussed, followed by the design methodology for the replication of the commercially manufactured horn. The design methodology of the basic pyramidal and conical horns operating in the Ku-band frequency range will also be discussed.

3.1 3D Printer

The Ultimaker 2+ is the only available printer for use within the UCT Electrical Engineering Department. Figure 3-1 shows the Ultimaker 2+. It is an FDM based printer with a build volume of 223 x 223 x 205 mm, that can print in PLA, ABS, nylon, and some other materials. Table 3-1 provides some specifications of the Ultimaker 2+ 3D printer.

Table 3-1: Specifications of the Ultimaker 2+ 3D printer.

Ultimaker 2+ specifications	
Dimensions	342 x 493 x 588 mm
Build volume	223 x 223 x 205 mm
Layer resolution	0.25 mm nozzle: 150 to 60 µm 0.40 mm nozzle: 200 to 20 µm
XYZ accuracy	12.5, 12.5, 5 µm
Build plate	50 °C to 100 °C heated glass
Nozzle temperature	180 °C to 260 °C
Materials	PLA, ABS, CPE, CPE+, PC, Nylon, TPU 95A

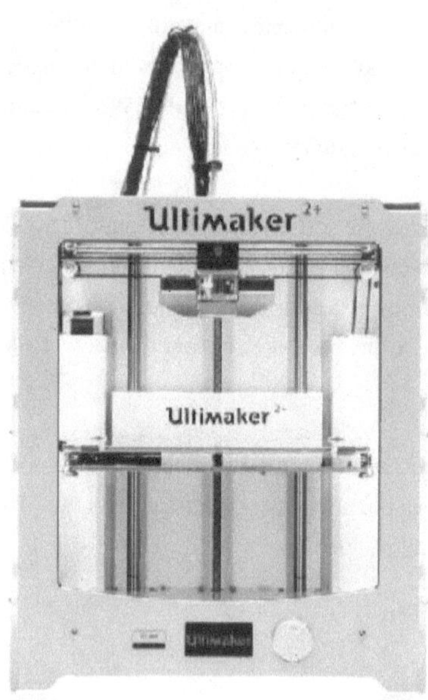

Figure 3-1: Image of the Ultimaker 2+ FDM based 3D printer [37].

The Ultimaker 2+ uses Cura as the print preparation software. It supports STL, OBJ, and AMF file types via upload through an SD card slot. Essentially, Cura is a 3D object slicer which takes a 3D object and slices it into layers and plans the path of the printer based on parameters specified by the user. The Department's Technical Officer manages a web interface called OctoPrint, which allows remote control and monitoring of the print job. The web interface also allows the user to control the bed and nozzle temperatures, view remotely the current print job via a camera, as well as use the G-code[1] viewer to view the layer by layer and planned path of any file that has been loaded onto the printer.

3.2 3D Printing Material

There were two materials available for 3D printing: polylactic acid (PLA) and acrylonitrile butadiene styrene (ABS). PLA has good tensile strength and surface quality, and is easy to work with at high speeds. However, they are not suitable for long term outdoor usage applications where the printed part will be exposed to temperatures higher than $50\,^{\circ}$C. Nonetheless, for the purposes of investigating the printing process of an antenna design, using PLA as a printing material is a good starting point since it is significantly easier to work with compared to ABS.

ABS has excellent mechanical properties and inter-layer adhesion in terms of printing using the Ultimaker 2+. It is not suitable for long term UV exposure as the properties of the ABS print could be negatively affected. The printed part should not be exposed to temperatures higher than $85\,^{\circ}$C. ABS is more complex to work with compared to PLA, but it is well-established as a plastic material used in the process of metallising plastics, which is a required step to make an antenna functional.

The full technical data sheet for the PLA and ABS used on the Ultimaker 2+ can be found in Appendices A and B. ABS is chosen as the material for printing the antennas designed in the following sections, because it can withstand higher temperatures than PLA, and it is a well-established material for utilisation in electroless and electrolytic plating.

---

[1]G-code is a numerical control programming language used by people to tell computerised machine tools how to make something.

3.3   X-Band Pyramidal Horn Antenna

In order to test the 3D printing and metallisation process, a commercially manufactured antenna is chosen to be replicated. One such antenna which was readily available is an X-band pyramidal horn antenna, operating between 8.2 to 12.4 GHz. The size of the antenna is within the specification of the 3D printer. The waveguide is a WR90 SMA (SubMiniature version A) right angle waveguide with operating frequency between 8.2 to 12.4 GHz, manufactured by Advanced Technical Materials Inc., part number 90-251-6, and serial number D267508-01. The specification sheet of the horn was not available, so the horn dimensions have been measured. The X-band horn measured dimensions are given in Table 3-2. The measured dimensions have been used to simulate the horn in FEKO.

Table 3-2: Measured dimensions of the X-band 8.2 to 12.4 GHz pyramidal horn.

Parameter	Value
Broad aperture length, A_H	38 mm
Narrow aperture length, A_E	34 mm
Flare length, L_E, L_H	29 mm
Waveguide length, L_g	30.5 mm
Waveguide broad length, a	22.9 mm
Waveguide narrow length, b	10.2 mm
Backshort length, l_b	8 mm
Probe length, l_p	6 mm
Probe radius, r_p	0.65 mm

To further evaluate the feasibility of using 3D printing and metallisation as a fabrication method, the designs of a basic pyramidal horn and conical horn antenna are considered in Sections 3.4 and 3.5.

3.4   Ku-Band Pyramidal Horn Antenna

In this section the procedure of designing the pyramidal horn antenna is described. The pyramidal horn design is based on the 3D printer size limitation. The frequency operation band chosen is the Ku-band, because it translates to an antenna size which is printable by the 3D printer. The frequency of operation is chosen around the centre of the operation band, 15 GHz. The horn dimensions are determined by using the theoretical design equations from Section 2.1.4.

For the operating frequency of 15 GHz, the wavelength is given as:

$$\lambda_\circ = \frac{c}{f_\circ}$$
$$= \frac{3 \times 10^8 \, \text{m/s}}{15 \times 10^9 \, \text{Hz}}$$
$$= 0.02 \, \text{m}$$

The dimensions of the waveguide are governed by the waveguide standard for the corresponding operating frequency. For 15 GHz in the Ku-band, the standard waveguide is WR62, with the broadside width of the waveguide, $a = 0.016\,\text{m}$ and the narrowside width, $b = 0.008\,\text{m}$. The waveguide wavelength is determined by Equations (2-10) and (2-11):

$$\lambda_g = \frac{\lambda_\circ}{\sqrt{1 - \left(\frac{\lambda_\circ}{2a}\right)^2}}$$
$$= \frac{0.02 \, \text{m}}{\sqrt{1 - \left(\frac{0.02 \, \text{m}}{2(0.016 \, \text{m})}\right)^2}}$$
$$= 0.026 \, \text{m}$$

Theoretically, a well-matched horn will satisfy the equations in Section 2.1.4 which determine the waveguide length, the probe length and the backshort length. Using Equations (2-12), (2-13) and (2-14), we can calculate the theoretical values for a well-matching horn:

$$L_g \geq \frac{3}{4}\lambda_g$$
$$\geq 0.019 \, \text{m}$$

The length of the waveguide is chosen to be 0.019 m. The probe length, l_p, and backshort length, l_b are calculated:

$$l_p = \frac{\lambda_\circ}{4}$$
$$= 0.005 \, \text{m}$$

$$l_b = \frac{\lambda_g}{4}$$
$$= 0.006\,\text{m}$$

The initial probe length and backshort length are chosen be to 5 mm and 6 mm respectively.

The 3D printer has a build volume of 223 mm $\times$ 223 mm $\times$ 205 mm. When 3D printing, distortion may occur due to material shrinkage, which causes the corners of the print to lift and detach from the build plate. Distortion usually occurs when the size of print encroaches the limits of the build plate. To ensure that this distortion is avoided, it has been decided to limit the maximum size of the antenna to 100 mm $\times$ 100 mm $\times$ 100 mm. Therefore, limiting the full length of the antenna to 100 mm. Using this as a guideline, the lengths of the aperture, L_E and L_H, have been chosen as 0.08 m which is approximately the limiting length less the waveguide length. Thus the aperture length in the E-field direction and the H-field direction can be calculated using Equations (2-8) and (2-9). The aperture length in the E-field direction is:

$$A_E = \sqrt{2\lambda_\circ L_E}$$
$$= \sqrt{2(0.02\,\text{m})(0.08\,\text{m})}$$
$$= 0.057\,\text{m}$$

and in the H-field direction:

$$A_H = \sqrt{3\lambda_\circ L_H}$$
$$= \sqrt{3(0.02\,\text{m})(0.08\,\text{m})}$$
$$= 0.069\,\text{m}$$

Table 3-3 summarises the calculated parameters of the pyramidal horn.

Table 3-3: Ku-band pyramidal horn theoretical parameters.

Ku-band pyramidal horn	
Parameter	**Value**
Frequency, $f_\circ$	15 GHz
Wavelength, $\lambda_\circ$	20 mm
Waveguide:	
Broadside width, a	16 mm
Narrowside width, b	8 mm
Length, L_g	19 mm
Wavelength, λ_g	25.6 mm
Flare:	
E-field aperture, A_E	57 mm
H-field aperture, A_H	69 mm
Length, L	81 mm
Probe length, l_p	5 mm
Backshort length, l_b	6.5 mm

The designed horn is modelled in FEKO as separate shapes, then merged, meshed, and inserting a waveguide port function on the waveguide. The Multi-level Fast Multiple Method (MLFMM) numerical method was used for the FEKO simulation. The MLFMM method enables the analysis of antennas in electrically large environments that allows near-field and far-field radiation pattern measurements. The antenna was simulated using different materials, such as a perfect conductor, copper, and nickel. However, the results were very identical, so only the results of the perfect conductor was shown. The optimisation tool in FEKO has been used on the backshort and probe length for optimum performance of the reflection coefficient characteristic. Table 3-4 lists the optimised backshort and probe lengths for the pyramidal horn. The theoretical backshort value is a mathematical approximation based on propagation in a vacuum, whereas the FEKO simulates propagation through air.

Table 3-4: FEKO-optimised parameters for the Ku-band pyramidal horn.

FEKO-optimised parameters	
Parameter	**Value**
Probe length, l_p	4.4 mm
Backshort length, l_b	4.8 mm

The steps taken in designing a pyramidal horn are relatively well defined in literature [14, 15, 17, 18]. The suggested theoretical guidelines have been used as the basis of the design, and although there has been a size restriction on the antenna due to the 3D printer capabilities, the focus of the project is to test the feasibility of the antenna fabrication method rather than the design process.

3.5 Ku-Band Conical Horn Antenna

In this section, a conical horn antenna is designed in the same frequency and using the same dimension limitation as the pyramidal horn to test the process of making another basic antenna of a different shape. The frequency of operation is chosen based on the same limitations as that of the pyramidal horn antenna, operating at 15 GHz. The conical horn is designed using the theoretical equations stated in Section 2.1.5.

Operating at 15 GHz, the wavelength of the horn is 20 mm. Firstly, the upper and lower cutoff frequencies of the conical horn is determined. The cutoff frequencies are governed by the waveguide diameter, d_g. The waveguide diameter is chosen such that the operating frequency is well within the lower $\left(\lambda_{c(low)}\right)$ and upper $\left(\lambda_{c(high)}\right)$ cutoff frequencies where only the fundamental mode exists. Using Equations (2-16) and (2-17), the waveguide diameter is chosen to be 0.013 m, giving a lower cutoff wavelength and frequency of:

$$
\begin{aligned}
\lambda_{c(low)} &= 1.706 d_g \\
&= 1.706 \times 0.013 \, \text{m} \\
&= 0.022 \, \text{m} \\
f_{c(low)} &= \frac{c}{\lambda_{c(low)}} \\
&= 13.57 \, \text{GHz}
\end{aligned}
$$

and an upper cutoff wavelength and frequency of:

$$
\begin{aligned}
\lambda_{c(high)} &= 1.306 d_g \\
&= 1.306 \times 0.013 \, \text{m} \\
&= 0.017 \, \text{m} \\
f_{c(high)} &= \frac{c}{\lambda_{c(high)}} \\
&= 17.67 \, \text{GHz}
\end{aligned}
$$

The wavelength of the waveguide can then be calculated using Equation (2-10):

$$\lambda_g = \frac{\lambda_o}{\sqrt{1 - \left(\frac{\lambda_o}{\lambda_c}\right)^2}}$$
$$= \frac{0.02\,\text{m}}{\sqrt{1 - \left(\frac{0.02\,\text{m}}{0.022\,\text{m}}\right)^2}}$$
$$= 0.046\,\text{m}$$

Using Equations (2-12), (2-13) and (2-14), we can calculate the theoretical values for a well-matching horn:

$$L_g \geq \frac{3}{4}\lambda_g$$
$$\geq 0.035\,\text{m}$$

The length of the waveguide is chosen to be $0.035\,\text{m}$. The probe length, l_p, and backshort length, l_b are calculated:

$$l_p = \frac{\lambda_o}{4}$$
$$= 0.005\,\text{m}$$
$$l_b = \frac{\lambda_g}{4}$$
$$= 0.009\,\text{m}$$

The initial probe length and backshort length are chosen be to $5\,\text{mm}$ and $9\,\text{mm}$ respectively. The flare diameter of the horn, D, is determined by the limitation of the print volume of the 3D printer and by using Equation (2-15). The length from the apex of the flare to the centre of the aperture, L, is chosen as $0.06\,\text{m}$, because it ensures that the overall horn shape is within the build volume of the the printer, and also that the slant of the horn is not too steep that it hinders the 3D printing process. The diameter of the horn is calculated:

$$D = \sqrt{3\lambda_o L}$$
$$= \sqrt{3(0.02\,\text{m})(0.06\,\text{m})}$$
$$= 0.06\,\text{m}$$

Once again, FEKO is used to simulate the horn, and the probe and backshort lengths have also been optimised for better reflection coefficient performance. Table 3-5 and 3-6 summarises the optimised design parameters of the conical horn antenna operating at 15 GHz.

Table 3-5: Ku-band conical horn theoretical parameters.

Ku-band conical horn	
Parameter	**Value**
Frequency, $f_\circ$	15 GHz
Wavelength, $\lambda_\circ$	20 mm
Waveguide:	
Diameter, d_g	13 mm
Length, L_g	34.7 mm
Wavelength, λ_g	46.3 mm
Flare:	
Diameter, D	60 mm
Length, L	60 mm
Probe length, l_p	5 mm
Backshort length, l_b	9 mm

Table 3-6: FEKO-optimised parameters for the Ku-band conical horn.

FEKO-optimised parameters	
Parameter	**Value**
Probe length, l_p	5.3 mm
Backshort length, l_b	9.6 mm

The design of the conical horn antenna follows the mathematical approximations given in literature. The simulation steps taken are similar to that of the pyramidal horn in Section 3.5, using the MLFMM numerical method in FEKO, as well as the optimisation tool for the backshort and probe lengths. Although some changes have been made in the probe and backshort parameters, the resulting design still creates a conical horn antenna which operates at the desired frequency, and falls within the 3D printer's printing capabilities.

In the next chapter, the fabrication of the X-band horn, the Ku-band pyramidal and conical horns will be described in detail, as well as the measurement of the X-band commercial horn and all fabricated horns.

Chapter 4

Fabrication and Measurement of Antenna Properties

In this chapter the fabrication of the X-band pyramidal horn antenna and Ku-band pyramidal and conical horn antennas will be discussed in detail. The measurement and testing of the fabricated antennas will also be discussed. All the antennas are fabricated using the FDM-based printing machine, Ultimaker 2+. The metallisation of the printed antennas is carried out by means of electroless and electrolytic plating at TraX or by painting. Measurement of the antenna properties are conducted using the anechoic chamber facilities and network analyser provided at Stellenbosch University (SU).

4.1   Fabrication

All antennas have been fabricated using the Ultimaker 2+ 3D printing machine. The parts of the antenna designs are modelled using SolidWorks Premium 2015, the software details are described in Section 1.4. SolidWorks allows the 3D models to be saved as an STL file which is the file type required by the Ultimaker 2+ Cura preparation software. The Cura software slices the 3D model into layers, and plans the path that the printer head will be taking to build the model layer by layer, from the bottom up. All the parts have been 3D printed using ABS material. As mentioned in Section 3.2, ABS has been chosen over PLA, because ABS has a higher temperature tolerance, and it is the most widely used material in metallisation processes, especially in electroplating plastics. All the antennas use coaxial connectors as the feed. SMA connectors have been chosen from RS components and the component drawing is shown in Appendix C.

4.2   Metallisation

Once the parts of the antenna are printed, they are metallised using one or both of two methods: plating, and painting. Metallisation by painting has been performed by applying a nickel-based aerosol paint on the surface of the parts. The product name is Nickel Screening Compound Plus (NSCP), product number ENSCP400H by Electrolube, purchased from RS Components, and the colour of the paint is dark grey in appearance. NSCP is an efficient EMI/RFI screening coating. The coating becomes touch dry 5 minutes after application, but maximum conductivity is only achieved after curing for 24 hours. The product is also available in bulk, but the aerosol form enables easier and more uniform application of the paint onto surfaces. The specification sheet of the paint can be found in Appendix D. The paint has been applied onto the parts in layers, allowing sufficient drying time before the next layer is applied.

Metallisation of the ABS plastic parts of the antenna has also been achieved by plating. This process has been performed by TraX Interconnect and their plating process follows similarly to the metallisation of ABS described in Section 2.4.1. First, electroless plating is performed on the ABS plastic parts of the 3D printed antenna, thereafter electrolytic plating is performed with copper as the plating material. Figure 4-1 illustrates the electroless plating station at TraX.

Figure 4-1: Electroless plating process used at TraX.

The station consists of 9 baths in which the plastic parts are submerged in a consecutive order starting from the bottom of Figure 4-1 to the top. The electroless plating process involves the following steps (one step per bath) in order:

- Cleanse: the plastic part is first cleansed to remove any foreign materials.

- Rinse: the part is rinsed to further remove any residue on the plastic.

- Rinse: a second rinse is given to thoroughly remove any residual dirt.

- Micro etching: this step is used to roughen the surface of the object to allow for better adhesion of the metal to the plastic. However, this step was omitted in the plating process for the antennas because it was agreed by the TraX chemists that the antenna surface did not require this step since the surface of the parts already have sufficient roughness.

- Rinse: first rinse after micro etching to wash off any unwanted substances (omitted).

- Rinse: second rinse (omitted).

- Pre-dip: this step involves producing a catalytic surface on the plastic part that is required to deposit palladium onto the surface.

- Catalyst: the surface of the plastic part is deposited with palladium which serves as the adhesive for the plastic part and the metal coating.

- Rinse: the plastic part is rinsed to prepare for the copper bath in the next step.

- Copper bath: the palladium metal previously deposited onto the surface of the plastic allows adhesion of the copper in the bath to the plastic surface. The thickness of the copper here is very thin, and is a preparation for the next process which is electrolytic plating.

- Rinse: the plastic part is given one final rinse.

The second plating process, called the electrolytic plating process, is performed to thicken the copper surface which is now already present on the surface of the plastic. This process only works on metallic objects, and is often used for coating metals with a different metal for specific applications or to prevent corrosion. Figures 4-2, 4-3 and 4-4 illustrate the electrolytic plating process used at TraX and the equipment involved in the plating process.

Figure 4-2: Electrolytic plating process used at TraX: plating station for electrolytic plating (left); and cleansing bath of the station (right).

Figure 4-3: Copper bath of the electrolytic plating process used at TraX.

Figure 4-4: Titanium rods (top) and copper spheres (bottom) used in the electroless plating process at TraX.

The plating procedure undertaken by TraX was a pilot experiment, therefore TraX did not bill for the plating service. In Sections 4.5 and 4.6 the outcome of the plating process will be discussed.

4.3 Measurement of Antenna Properties

After fabrication and metallisation of the antennas, the properties of the antenna are measured using the anechoic chamber facilities at SU. The measured properties of the antenna are the reflection coefficient, the gain, and the radiation pattern.

4.3.1 Reflection Coefficient

The reflection coefficient of the antenna is measured using an Agilent N5242A PNA-X Microwave Network Analyser which has a maximum operation frequency of 26.5 GHz. This property has been measured by first calibrating the port to remove errors in the cable and connectors, also known as error correction. The parameter S_{11} is then measured and recorded on the network analyser. The S_{11} parameter is a measure of the reflection coefficient and determines the ratio of how much power is reflected by the antenna versus how much is transmitted.

4.3.2 Gain

The gain of the antenna is measured using the three-antenna method. This method can be used when there is no standard gain horn available as a reference antenna for measuring the antenna under test (AUT). The three-antenna method is based on the Friis transmission equation (2-6):

$$P_R = \frac{P_T G_T G_R \lambda_\circ^2}{(4\pi R)^2} \quad [\text{W}]$$

Another two antennas of unknown gain can be used to perform this measurement. The AUT (x) and the other two antennas (y and z) are configured and measured in the following manner:

- Place one antenna as a transmitter and another as a receiver by a separation distance where both antennas are in the far field region.

- Measure the distance between the two antennas.

- Calibrate the antennas using the network analyser such that the measured S_{21} parameter is the effective power transmission ratio, i.e., any losses in the cables are already accounted for in the S_{21} measurement.

- Record the transmission coefficient S_{21} data, which is equivalent to the power transmission ratio P_R/P_T.

- Repeat with different combination of antennas until three sets of the data is recorded.

The Friss equation can be rearranged to correspond with the configuration:

$$S_{xy} = \frac{P_x}{P_y} = \frac{G_x G_y \lambda_\circ^2}{(4\pi R_{xy})^2} \tag{4-1}$$

$$S_{xz} = \frac{P_x}{P_z} = \frac{G_x G_z \lambda_\circ^2}{(4\pi R_{xz})^2} \tag{4-2}$$

$$S_{yz} = \frac{P_y}{P_z} = \frac{G_y G_z \lambda_\circ^2}{(4\pi R_{yz})^2} \tag{4-3}$$

With the measured data, the three unknown gains can be found by solving the simultaneous equations, however, only the gain of the AUT is of interest.

4.3.3 Radiation Pattern

The radiation pattern of the antenna is measured in the anechoic chamber by placing the AUT as a receiver on a pedestal, and another antenna which operates at the same centre frequency is used as the transmitter. The pedestal is rotated in the elevation and azimuth planes and the power received at every degree of rotation is measured and translated to a full three dimensional radiation pattern of the AUT. The azimuth and elevation cuts of the radiation pattern for co-pol and x-pol measurements can be extracted from the raw data of the full 3D radiation pattern measurement. Figure 4-5 demonstrates the set-up of the radiation pattern measurement in the anechoic chamber.

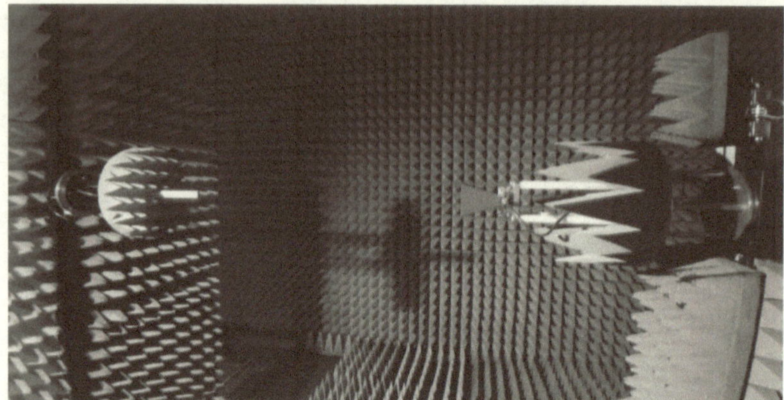

Figure 4-5: Painted pyramidal horn mounted on the pedestal (right side of the photograph) in the anechoic chamber at SU for measurement.

Generally the radiation pattern of an antenna is measured in the far field region, but in this anechoic chamber, the pattern of the antenna can be measured in the near field region, and the measured pattern can be transformed into a far field pattern by the measurement software[1] used in the anechoic chamber. The software measures the antenna pattern in the near field region and converts the measurements by a Fourier transform into the far field result. The software records the raw data and allows the plot of the radiation pattern in the selected plane for a quick analysis. All the radiation patterns of the antennas have been recorded, the raw data are saved and can be further analysed using a more common programming language such as Matlab.

4.4 X-Band Pyramidal Horn

The X-band pyramidal horn in Section 3.3 is fabricated by 3D printing the antenna, and the antenna is then metallised using the paint method described in Section 4.2. The paint used is the aerosol form of NSCP.

The 3D model of the X-band pyramidal horn is modelled as a replication of the commercial horn antenna, where the horn and waveguide are two separate parts. The waveguide is further modelled as two separate parts to simplify the metallisation process. The three parts are fitted together by

---

[1]NSI Near Field Systems inc. NSI2000 Antenna Measurement System v4.13.45

using bolts. The antenna feed uses the coaxial SMA connector, and is secured onto the waveguide using bolts. To ensure better guided wave propagation, a brass bush has been fabricated to fit between the dielectric of the connector and the waveguide wall. Figure 4-6 shows a photograph of the commercial X-band pyramidal horn and Figure 4-7 shows the fabricated X-band pyramidal horn.

Figure 4-6: Photograph of commercial X-band pyramidal horn.

Figure 4-7: Photograph of fabricated X-band pyramidal horn.

The properties of the fabricated X-band horn as well as the X-band commercial horn are then measured in the anechoic chamber facility at SU. The reflection coefficient of the horns are measured using the network analyser. A full radiation pattern has been measured from 8.2 to 12.4 GHz. The gain of the AUTs are measured using the three-antenna method described in Section 4.3. Two other antennas of unknown gain, operating in the X-band are used: an X-band WR90 open-ended waveguide (OEWG), and an AEL (American Electronic Laboratories) 2 to 18 GHz double-ridged rectangular horn. These two antennas were chosen to avoid using more than one AUT in the three antenna measurement, as they are antennas known to operate in the desired frequency range. Table 4-1 summarises the measured transmission coefficient and separation distance of each antenna couple configuration.

Table 4-1: Summary of parameters measured for the fabricated X-band pyramidal horn used in the gain measurement.

Antenna 1	Antenna 2	S_{21} at 10 GHz	Distance, R
Fabricated X-band horn	WR90 OEWG	-32.5 dB	0.625 m
Fabricated X-band horn	AEL horn	-25.9 dB	0.460 m
WR90 OEWG	AEL horn	-29.4 dB	0.503 m

Antenna x is the fabricated X-band pyramidal horn, antenna y is the WR90 OEWG, and antenna z is the AEL 2 to 18 GHz horn. Table 4-2 summarises the measured S_{21} and separation distances of the commercial X-band horn gain measurement configuration.

Table 4-2: Summary of parameters measured for the commercial X-band pyramidal horn used in the gain measurement.

Antenna 1	Antenna 2	S_{21} at 10 GHz	Distance, R
Commercial X-band horn	WR90 OEWG	-35.5 dB	1.110 m
Commercial X-band horn	AEL horn	-29.8 dB	0.946 m
WR90 OEWG	AEL horn	-35.2 dB	0.987 m

Antenna x is the commercial X-band pyramidal horn, antenna y is the WR90 OEWG, and antenna z is the AEL 2 to 18 GHz horn. A mount interface was required for the antenna to be mounted onto the pedestal in the anechoic chamber. This interface was designed and then fabricated using the 3D printer, and the design is shown in Appendix E.

4.5   Ku-Band Pyramidal Horn

The Ku-band pyramidal horn in Section 3.4 is fabricated by 3D printing the antenna, and metallised using the two methods described in Section 4.2: copper plating performed by TraX, and painting with NSCP. The copper-plated antenna is shown in Figure 4-8.

Figure 4-8: Photograph of copper Ku-band pyramidal horn plated by TraX Interconnect.

The plating process performed by TraX has been a first-time experiment for them. It is evident that the antenna parts have not been completely plated over the entire surface. The discontinuation of the plating is only seen on the edges of the parts. A possible reason for this phenomenon may be due to air bubbles which could have formed on these edges during the electroless plating process. In order to remedy the effect that the discontinuations or "plating holes" may have on the efficiency of the wave propagation through the antenna, copper tape has been used to mend these holes. The copper tape used is a Hi-Bond HB 720A Conductive Copper Tape purchased from RS Components. The tape has a foil thickness of 0.025 mm and uses conductive acrylic adhesive. Figure 4-9 shows the Ku-band pyramidal horn metallised using the nickel-based paint.

Figure 4-9: Photograph of Ku-band pyramidal horn painted with nickel-based EMI shielding spray paint.

The two Ku-band pyramidal horns fabricated using two different metallisation methods have been tested in the anechoic chamber. The reflection coefficient, radiation patterns and gain of the Ku-band pyramidal horns have been measured following the same procedure as that of the X-band pyramidal horn, described in Section 4.3. Table 4-3 summarises the measured transmission coefficient and separation distance of each antenna couple configuration for the gain measurement of the painted Ku-band pyramidal horn.

Table 4-3: Summary of parameters measured for the painted Ku-band pyramidal horn used in the gain measurement.

Antenna 1	Antenna 2	S_{21} at 15 GHz	Distance, R
Painted Ku-band pyramidal horn	WR62 OEWG	$-30.4\,$dB	$0.731\,$m
Painted Ku-band pyramidal horn	AEL Horn	$-24.7\,$dB	$0.646\,$m
WR62 OEWG	AEL Horn	$-37.8\,$dB	$0.726\,$m

Antenna x is the painted Ku-band pyramidal horn, antenna y is the WR62 OEWG, and antenna z is the AEL 2 to 18 GHz horn. Table 4-4 summarises the measured S_{21} and separation distances of the copper-plated Ku-band horn gain measurement configuration.

Table 4-4: Summary of parameters measured for the copper-plated Ku-band pyramidal horn used in the gain measurement.

Antenna 1	Antenna 2	S_{21} at 15 GHz	Distance, R
Plated Ku-band pyramidal horn	WR62 OEWG	−27.2 dB	0.585 m
Plated Ku-band pyramidal horn	WR90 OEWG	−25.1 dB	0.587 m
WR62 OEWG	WR90 OEWG	−37.6 dB	0.667 m

Antenna x is the plated Ku-band pyramidal horn, antenna y is the WR62 OEWG, and antenna z is the WR90 OEWG. A mount interface has also been used for the Ku-band pyramidal horns, and the design of the interface is shown in Appendix E.

4.6 Ku-Band Conical Horn

The Ku-band conical horn in Section 3.5 has been fabricated and also metallised using two methods, by plating and painting. The copper plating has been performed by TraX following the same method described in Section 4.2. Figure 4-10 and 4-11 respectively show the Ku-band conical horn copper-plated by TraX, and painted with NSCP.

Figure 4-10: Photograph of copper Ku-band conical horn plated by TraX.

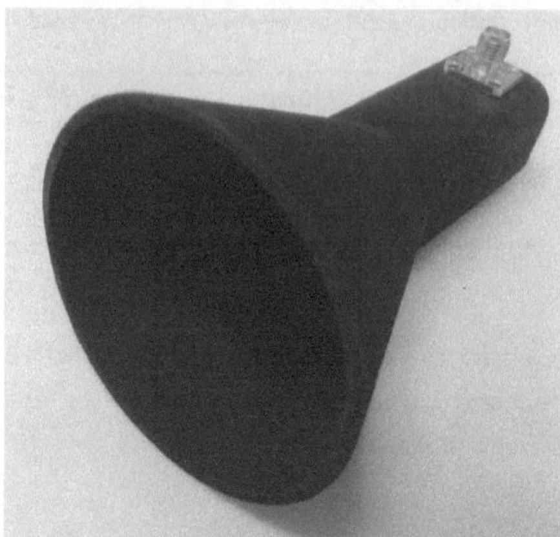

Figure 4-11: Photograph of Ku-band conical horn painted with NSCP.

The surface of the copper-plated conical horn has less plating holes compared to that of the copper-plated pyramidal horn. This further supports the possibility that the holes are a result of air bubbles formed on the surface of the parts during the electroless plating process. The antenna properties of the metallised conical horns have been measured in the anechoic chamber following the same procedures as the X-band and Ku-band pyramidal horns. Table 4-5 summarises the measured transmission coefficient and separation distance of each antenna couple configuration for the gain measurement of the painted Ku-band conical horn.

Table 4-5: Summary of parameters measured for the painted Ku-band conical horn used in the gain measurement.

Antenna 1	Antenna 2	S_{21} at 15 GHz	Distance, R
Painted Ku-band conical horn	WR62 OEWG	−35.8 dB	0.738 m
Painted Ku-band conical horn	AEL horn	−30.4 dB	0.653 m
WR62 OEWG	AEL horn	−37.8 dB	0.726 m

Antenna x is the painted Ku-band conical horn, antenna y is the WR62 OEWG, and antenna z is the AEL 2 to 18 GHz horn. Table 4-6 summarises the measured S_{21} and separation distances of the copper-plated Ku-band conical horn gain measurement configuration.

Table 4-6: Summary of parameters measured for the copper-plated Ku-band conical horn used in the gain measurement

Antenna 1	Antenna 2	S_{21} at 15 GHz	Distance, R
Plated Ku-band conical horn	WR62 OEWG	−28.5 dB	0.592 m
Plated Ku-band conical horn	WR90 OEWG	−26.4 dB	0.594 m
WR62 OEWG	WR90 OEWG	−37.6 dB	0.667 m

Antenna x is the plated Ku-band conical horn, antenna y is the WR62 OEWG, and antenna z is the WR90 OEWG. The mount interface used for the conical horns is shown in Appendix E.

4.7 Cost Analysis

This section will describe and list all costs incurred in fabricating the antennas from 3D printing to metallising.

Table 4-7 lists all parts that have been 3D printed using the Ultimaker 2+.

Table 4-7: 3D printing costs of all parts printed using the Ultimaker 2+.

Part(s)	Filament used (mm)	Filament weight used (grams)	Total cost (ZAR)
X-band pyramidal horn (flare, waveguide, endpiece)	3335	8.34	100.54
Ku-band pyramidal horn (flare and waveguide)	5674	14.19	173.08
Ku-band pyramidal horn (endpiece)	229	0.57	7.91
Ku-band conical horn (flare and waveguide)	3569	8.93	107.62
Ku-band pyramidal horn coupling mount	7700	19.26	175.20
Ku-band conical horn coupling mount	7212	18.04	165.14
Ku-band pyramidal and conical horn coupling clasp	2139	5.35	63.06
Ku-band conical horn coupling clasp plate	323	0.81	8.77
Total	**30181**	**75.49**	**801.32**

The total 3D printing costs amounted to ZAR 801.32. The SMA connectors purchased from RS Components cost ZAR 115.93 each, amounting to ZAR 579.65. The brass bushes used for the SMA connector transition for all the antennas have been manufactured by ACUFAB and amounted to ZAR 240. The copper tape and NSCP purchased from RS Components cost ZAR 297.77 and ZAR 727.94 per unit respectively. Approximately one unit of the NSCP has been used for the fabricated antennas, and approximately 2 m of the 33 m long copper tape has been used, amounting to ZAR 18.05. The plating work performed by TraX has not been billed since they consider the plating process as an experiment for them. Table 4-8 summarises the costs of all materials used in the fabrication and metallisation of the five antennas: the X-band pyramidal horn, the plated and painted Ku-band pyramidal and conical horns.

Table 4-8: Summary of all costs incurred in the fabrication, metallisation, and measurement of the five antennas: the X-band pyramidal horn, the plated and painted Ku-band pyramidal and conical horns.

Item	Incurred cost (ZAR)
3D printing all parts	801.32
SMA connectors	579.65
Brass bushes	240
Copper tape	18.05
Conductive paint	727.94
Total	**2366.96**

Each antenna costs approximately ZAR 475 to fabricate, metallise and measure. Traditionally, such horn antennas of higher operating frequency are manufactured using subtractive manufacturing methods such as CNC machining. In recent years it has also become possible to manufacture metallic parts using additive manufacturing methods such as SLM. A company located in Johannesburg, South Africa provides such a service, 3D printing in metal. They quoted approximately ZAR 4500 (excluding VAT) for the fabrication of one unit of the Ku-band pyramidal horn using AlSi10Mg material. This is significantly more expensive than the method conducted in this project. Since one of the requirements of the project is to validate the process of fabricating antennas in a low-cost manner, the option of 3D printing in metal has not been pursued.

Chapter 5

Measured Antenna Properties

This section presents all the measured antenna properties for the X-band pyramidal horn, the Ku-band pyramidal horn, and the Ku-band conical horn. All the raw data from measurements at SU, on the network analysers, and extraction from FEKO are in comma-separated values (.csv) format. The processing of raw data and plotting of graphs are all performed on Matlab v2018a for easier analysis and calculation of antenna properties. The plots presented are the reflection coefficient measurements, the gain measurements, and the azimuth and elevation co- and x-pol radiation patterns. The gain used for the FEKO measurements is the realised gain. The measured antenna properties are compared with the FEKO simulated measurements and are discussed in further detail.

5.1   X-Band Pyramidal Horn

The measurements for three X-band pyramidal horns are presented in this section: the commercial horn, the FEKO simulated horn based on the dimensions of the commercial horn, and the 3D printed and spray-painted horn which is also based on the commercial horn for comparison purposes.

5.1.1 Reflection Coefficient of X-Band Pyramidal Horn

Figure 5-1 shows the reflection coefficient of the FEKO simulated X-band horn, the reflection coefficient of the commercial X-band horn, and the reflection coefficient of the painted X-band horn from 8 to 12 GHz. The respective reflection coefficient values at 10 GHz are −10.4 dB, −25.4 dB, and −11.9 dB.

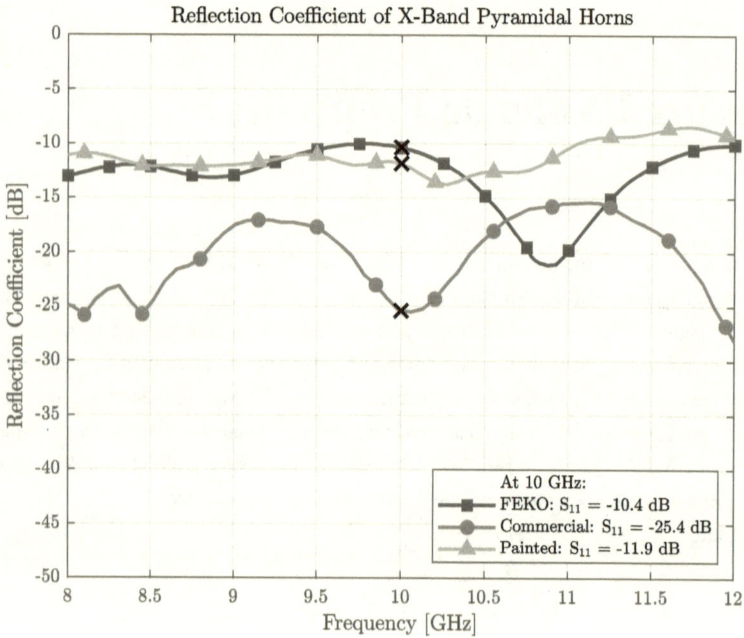

Figure 5-1: Reflection coefficient of X-band horns for FEKO simulation, commercial horn, and painted horn: the 'x' markers indicate the S_{11} at 10 GHz and the values are displayed in the legend.

Both the FEKO simulation and the painted horn have been based on the physical dimensions of the commercial horn. The commercial horn has a better response across the majority of the band and has a better overall response throughout the frequency band for which it has been designed (8.2 to 12.4 GHz). The commercial horn achieves a response of less than −14 dB across the entire operating frequency band. The painted horn achieves a response of less than −8 dB across the entire operating frequency band, and less than −11 dB at 10 GHz.

5.1.2   Gain of X-Band Pyramidal Horn

Antenna gain measures the ability of an antenna to direct energy in a particular direction or pattern. This is an important property of an antenna to determine how efficiently it radiates energy. Figure 5-2 shows the gain of the FEKO simulated X-band horn, the commercial X-band horn, and the painted X-band horn from 8 to 12 GHz. The respective gain values at 10 GHz are 11.2 dBi, 8.4 dBi, and 9.2 dBi.

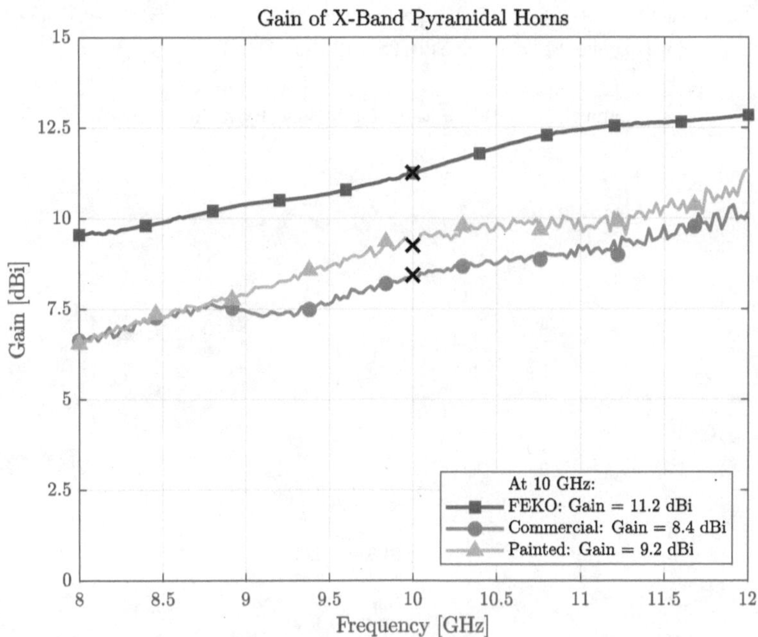

Figure 5-2: Gain of X-band horns for FEKO simulation, commercial horn, and painted horn: the 'x' markers indicate the gain at 10 GHz and the values are displayed in the legend.

The measured gain of the commercial and painted horns are less than that of the FEKO simulation. The value is lower because the commercial and painted horns are fabricated horns which will have losses due to the material used. Furthermore, the FEKO simulation determines the gain using optimal parameters: the material used in simulation is that of a perfect lossless conductor.

5.1.3 Radiation Pattern of X-Band Pyramidal Horn

The radiation pattern of an antenna shows in which direction the input power is radiated. Some important parameters determined from the radiation pattern, are the HPBW which indicates how concentrated the radiated energy is in the main beam direction (−3 dB from the peak), and the SLL which further characterises the radiation pattern. Generally, a low SLL is desired since this means less power is radiated in the direction of the side of the antennas. Figure 5-3 shows the azimuth radiation pattern for the FEKO simulated X-band horn, the commercial X-band horn, and the painted X-band horn at 10 GHz. The respective azimuth HPBWs are 50.4°, 49.7° and 46.4°, while the respective azimuth SLLs are 25.5 dB, 25.0 dB and 23.0 dB.

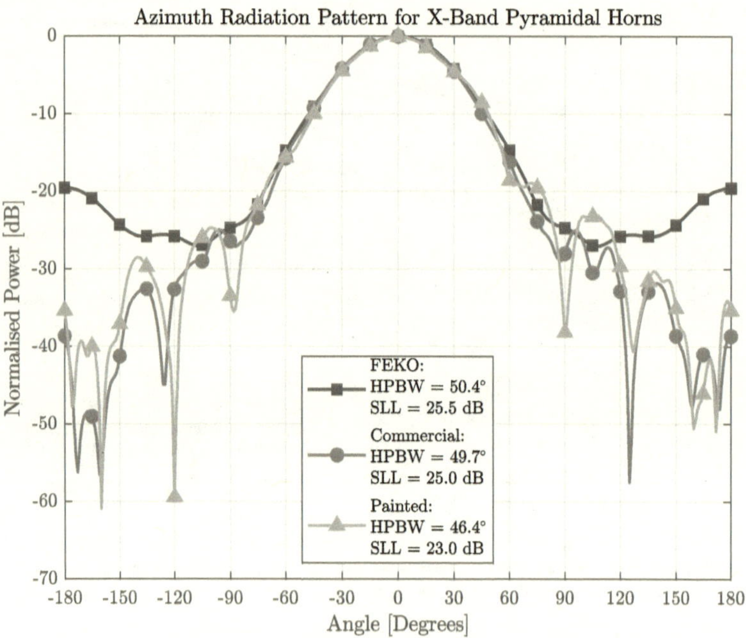

Figure 5-3: Azimuth radiation pattern of the X-band horns for FEKO simulation, commercial horn, and painted horn.

Figure 5-4 shows the elevation radiation pattern for the FEKO simulated X-band horn, the measured pattern for the commercial X-band horn, and the measured pattern for the painted X-band horn at 10 GHz. The respective elevation HPBWs are 45.7°, 44.3° and 41.4°, while the respective elevation SLLs are 19.7 dB, 19.3 dB and 16.6 dB. There is a backlobe difference of approximately 15 dB, this difference may be due to the physical configuration in the anechoic chamber as shown in Figure 4-5. On the far right in Figure 4-5, the pedestal upon which the AUT is mounted has absorbers positioned to avoid back reflections. However, these absorbers can interfere with the backlobe measurements, resulting in the measured backlobe to be much lower than that of the FEKO measurement.

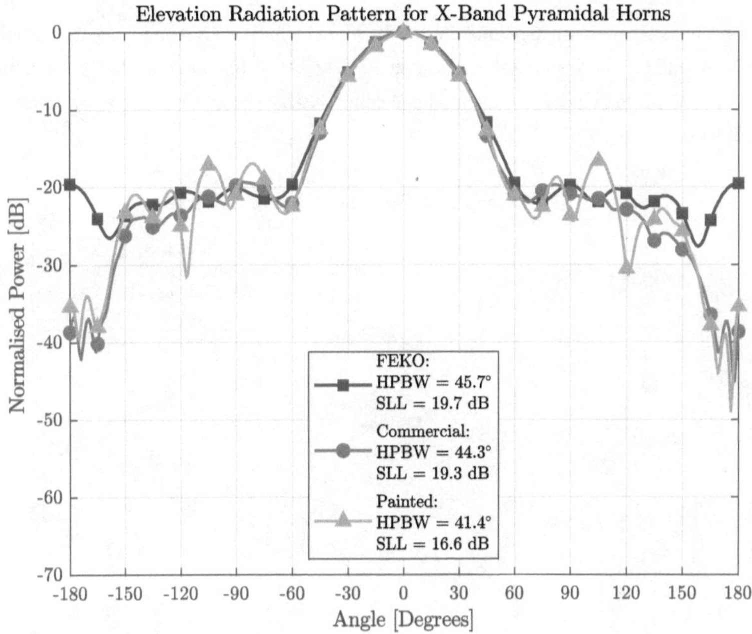

Figure 5-4: Elevation radiation pattern of the X-band horns for FEKO simulation, commercial horn, and painted horn.

The measured elevation HPBW of the horn compares well with the FEKO simulation, and the SLLs are acceptable, with the highest SLL at 16.6 dB for the painted horn.

5.1.4 Cross-Polarisation Pattern of X-Band Pyramidal Horn

The x-pol measurement of an antenna determines how much power is radiated from one antenna to another antenna when their polarisations are orthogonal to each other. Linearly polarised antennas are generally designed for optimal performance in the co-pol configuration, therefore it is desired for the x-pol measurement to be minimal.

Figure 5-5 shows the azimuth x-pol pattern for the FEKO simulated X-band horn, the commercial X-band horn, and the fabricated X-band horn at 10 GHz. At 0° line of sight (LoS), the respective measured power for the FEKO, commercial and painted X-band horns are −69.8 dB, −36.9 dB and −36.8 dB.

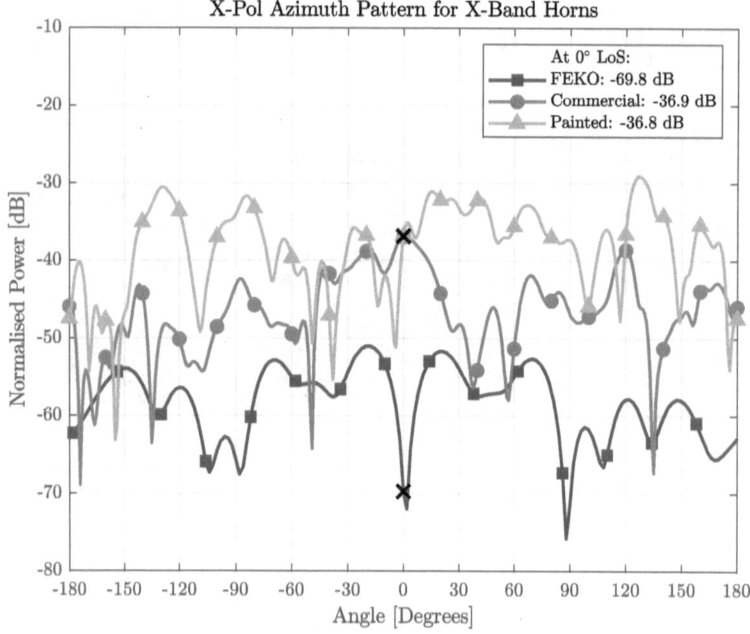

Figure 5-5: Azimuth x-pol pattern of the X-band horns for FEKO simulation, commercial horn, and painted horn: the 'x' markers indicate the normalised power at 0° LoS and the values are displayed in the legend.

Figure 5-6 shows the elevation x-pol radiation pattern for the FEKO simulated X-band horn, the commercial X-band horn, and the painted X-band horn at 10 GHz. At 0° line of sight (LoS), the respective measured power for the FEKO, commercial and painted X-band horns are −58.6 dB, −36.9 dB and −36.8 dB.

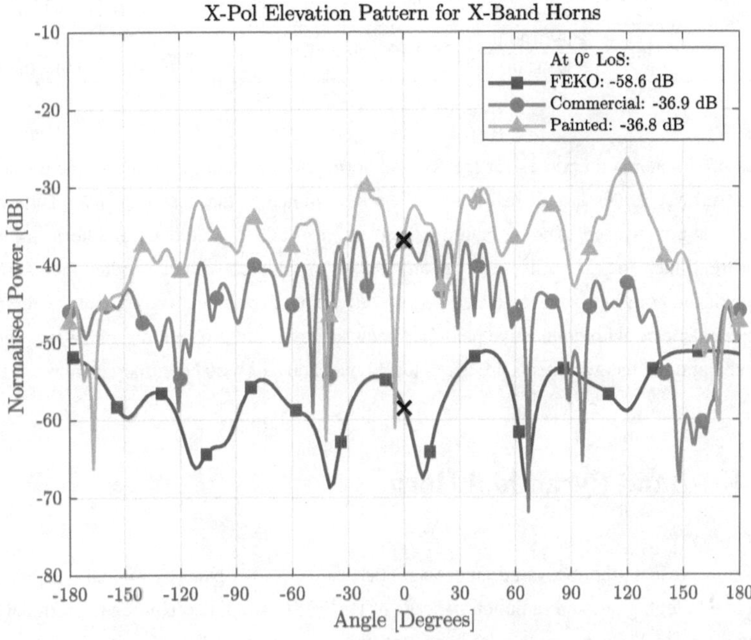

Figure 5-6: Elevation x-pol pattern of the X-band horns for FEKO simulation, commercial horn, and painted horn: the 'x' markers indicate the normalised power at 0° LoS and the values are displayed in the legend.

The fundamental mode of propagation is TE_{10} where the E-field distribution is only in the vertical direction, and therefore has a low x-pol leakage.

5.1.5 Summary

Table 5-1 presents a summary of all the measured antenna properties for the X-band pyramidal horn, namely the X-band commercial horn, the 3D printed and spray-painted X-band horn, and the FEKO simulated X-band horn based on the commercial horn.

68

Table 5-1: Summary of measured antenna properties for the X-band pyramidal horns at 10 GHz.

		S_{11}	Gain	HPBW	SLL	X-Pol
FEKO	Azimuth	−10.4 dB	11.2 dBi	50.4°	25.5 dB	−69.8 dB
	Elevation			45.7°	19.7 dB	−58.6 dB
Commercial Horn	Azimuth	−25.4 dB	8.4 dBi	49.7°	25.0 dB	−36.9 dB
	Elevation			44.3°	19.3 dB	−36.9 dB
Painted Horn	Azimuth	−11.9 dB	9.2 dBi	46.4°	23.0 dB	−36.8 dB
	Elevation			41.4°	16.6 dB	−36.8 dB

The measured antenna properties for the X-band horns matches that of simulated measurements. The achieved S_{11} values are all below −10 dB, which indicates that less than 1 % of the power transmitted is not radiated. The measured gain from 8 to 12 GHz for the X-band horns all follow a similar trendline. The azimuth and elevation HPBWs match well with each other, indicating that the shape of the horn has been replicated successfully. This experiment proves that the replicated horn antenna coated with conductive paint has been successful in propagating EM waves through the antenna and the measurements are very closely matched to a metal commercial horn.

5.2 Ku-Band Pyramidal Horn

This section presents the measured antenna properties of the Ku-band pyramidal horn. The reflection coefficient, gain, and radiation patterns of the FEKO simulated Ku-band pyramidal horn, the copper-coated Ku-band pyramidal horn, and the spray-painted Ku-band pyramidal horn are presented.

5.2.1 Reflection Coefficient of Ku-Band Pyramidal Horn

Figure 5-7 shows the reflection coefficient of the FEKO simulated Ku-band pyramidal horn, the copper pyramidal horn, and the painted pyramidal horn from 14 to 16 GHz. The respective reflection coefficient values at 15 GHz are −51.2 dB, −23.3 dB, and −24.2 dB.

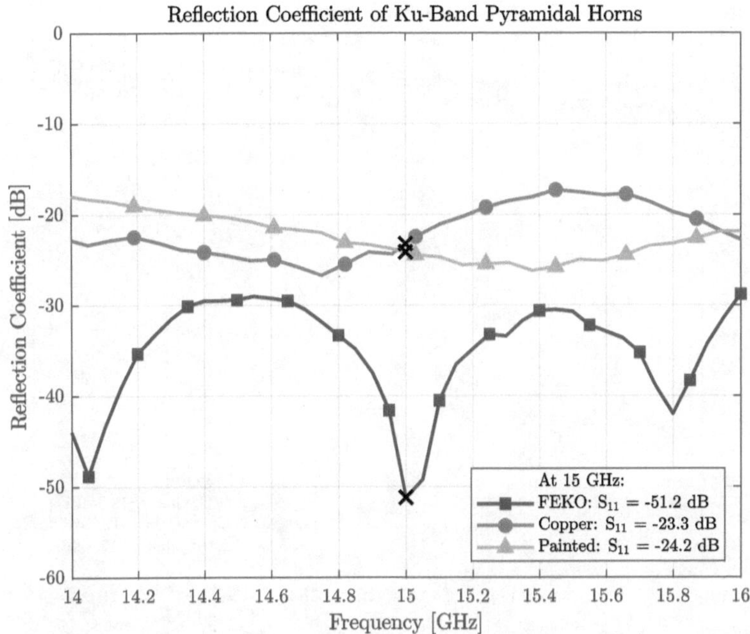

Figure 5-7: Reflection coefficient of Ku-band pyramidal horns for FEKO simulation, copper horn, and painted horn.

The reflection coefficient achieved by the fabricated Ku-band horns are less than -23 dB at the desired operating frequency of 15 GHz. This means that only less than 0.5 % of the power transmitted is lost in transmission. It is also observed that over the entire measured frequency band between 14 and 16 GHz, the reflection coefficient is well below -15 dB, indicating that the designed antenna can operate over a very wide bandwidth.

5.2.2 Gain of Ku-Band Pyramidal Horn

Figure 5-8 shows the gain of the FEKO simulated Ku-band pyramidal horn, the copper pyramidal horn, and the painted pyramidal horn from 14 to 16 GHz. The respective gain values at 15 GHz are 18.8 dBi, 17.7 dBi, and 17.5 dBi.

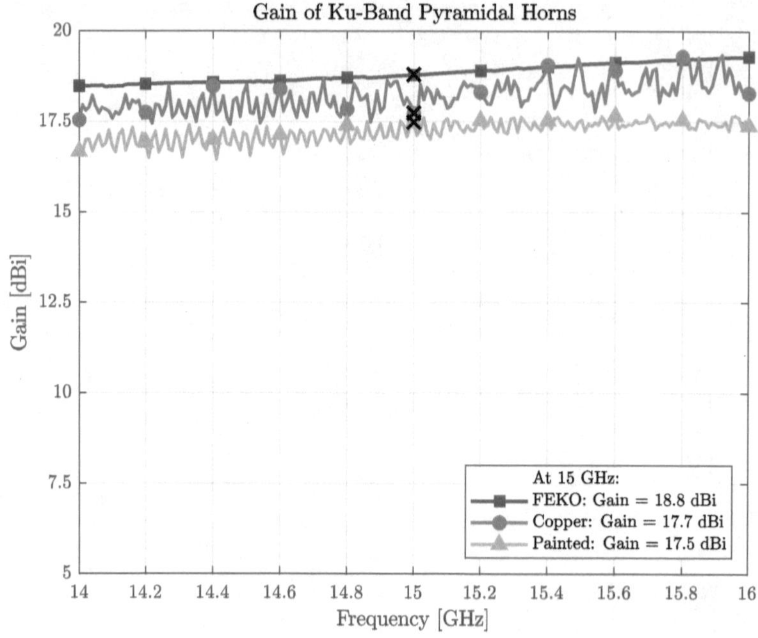

Figure 5-8: Gain of Ku-band pyramidal horns for FEKO simulation, copper horn, and painted horn.

The measured gain values of the horn match very closely with the simulated value, and all three measurements follow a similar trendline over the measured frequency range between 14 and 16 GHz.

5.2.3 Radiation Pattern of Ku-Band Pyramidal Horn

Figure 5-9 shows the azimuth radiation pattern for the FEKO simulated Ku-band pyramidal horn, the Ku-band copper pyramidal horn, and the Ku-band painted horn at 15 GHz. The respective azimuth HPBWs are 20.8°, 21.3° and 21.0°, while the respective azimuth SLLs are 29.3 dB, 26.8 dB and 27.3 dB.

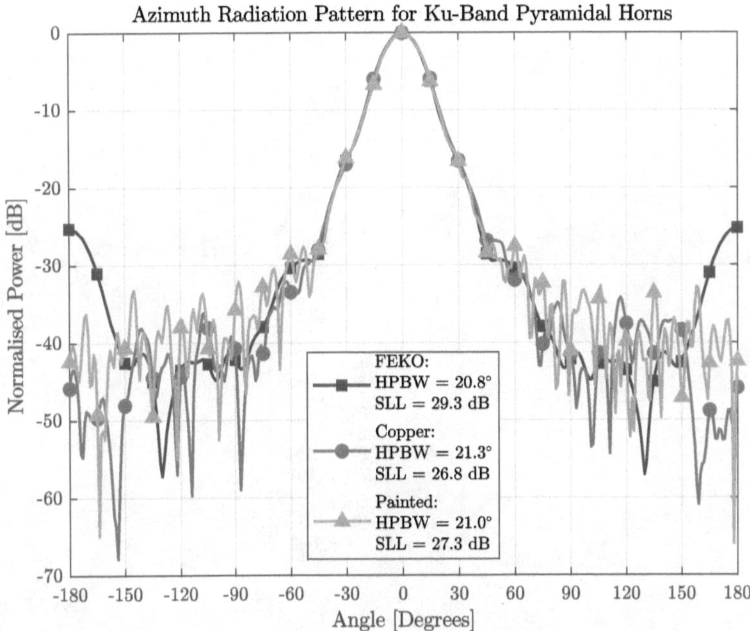

Figure 5-9: Azimuth radiation pattern of Ku-band pyramidal horns for FEKO simulation, copper horn, and painted horn.

The elevation radiation pattern of the Ku-band pyramidal horns are shown in Figure 5-10. It shows the elevation radiation pattern for the FEKO simulated Ku-band pyramidal horn, the Ku-band copper pyramidal horn, and the Ku-band painted horn at 15 GHz. The respective elevation HPBWs are 18.8°, 19.2° and 18.4°, while the respective elevation SLLs are 10.7 dB, 10.5 dB and 9.9 dB.

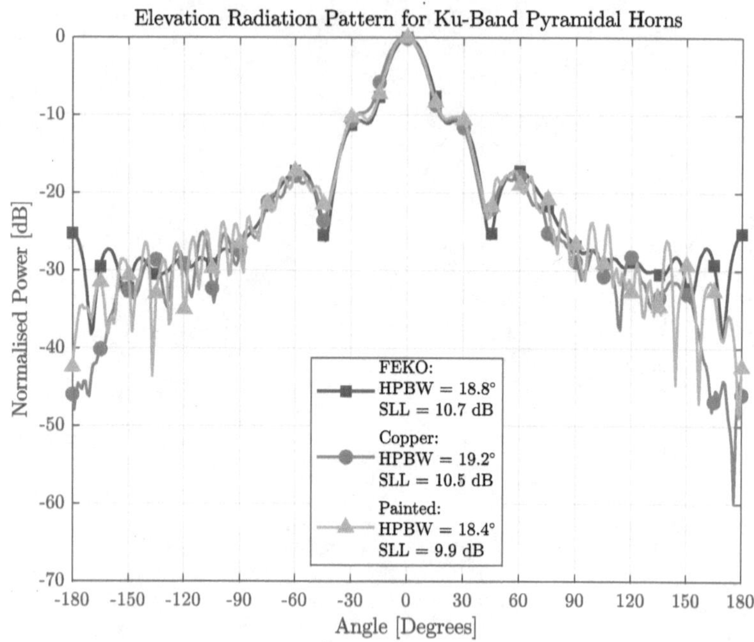

Figure 5-10: Elevation radiation pattern of Ku-band pyramidal horns for FEKO simulation, copper horn, and painted horn.

From the overlay of the three plots, it can be seen that the main beam matches very well. The minimal variations in the sidelobes are neglectable since the level of the sidelobes are low.

5.2.4 Cross-Polarisation Pattern of Ku-Band Pyramidal Horn

The x-pol patterns for the Ku-band pyramidal horns are presented. Figure 5-11 shows the azimuth x-pol pattern for the FEKO simulated Ku-band pyramidal horn, the copper pyramidal horn, and the painted horn at 15 GHz. It is shown in Figure 5-11 that the x-pol measurement is significantly lower in the simulated measurement. At 0° LoS, the measured azimuth x-pol pattern shows a relative power level of −38.2 dB for the copper-coated horn, and −32.2 dB for the painted horn.

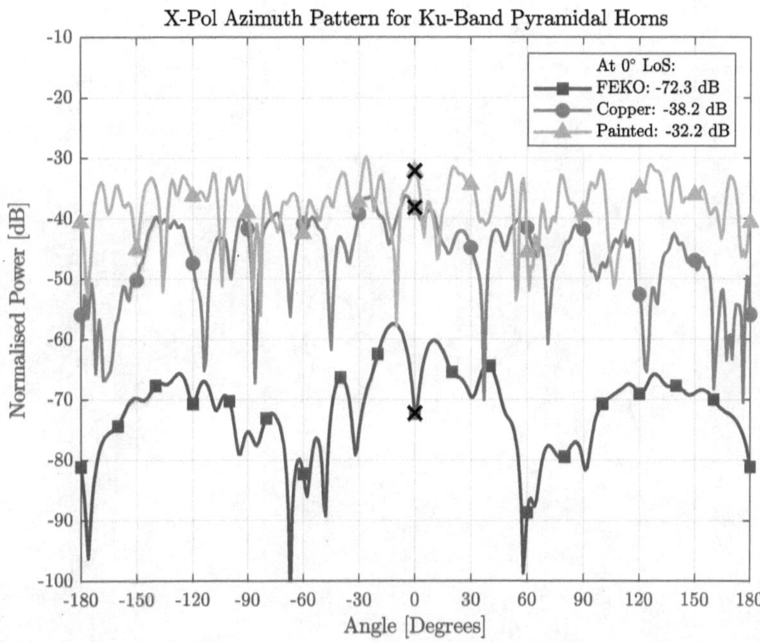

Figure 5-11: Azimuth x-pol pattern for Ku-band pyramidal horns for FEKO simulation, copper horn, and painted horn.

Figure 5-12 shows the elevation x-pol pattern for the FEKO simulated Ku-band pyramidal horn, the copper pyramidal horn, and the painted horn at 15 GHz. Once again, it is evident that the simulated measurement is significantly lower than that of the fabricated horns. At 0° LoS, the copper-coated horn has a power level of -38.2 dB and the painted horn has power level of -32.2 dB.

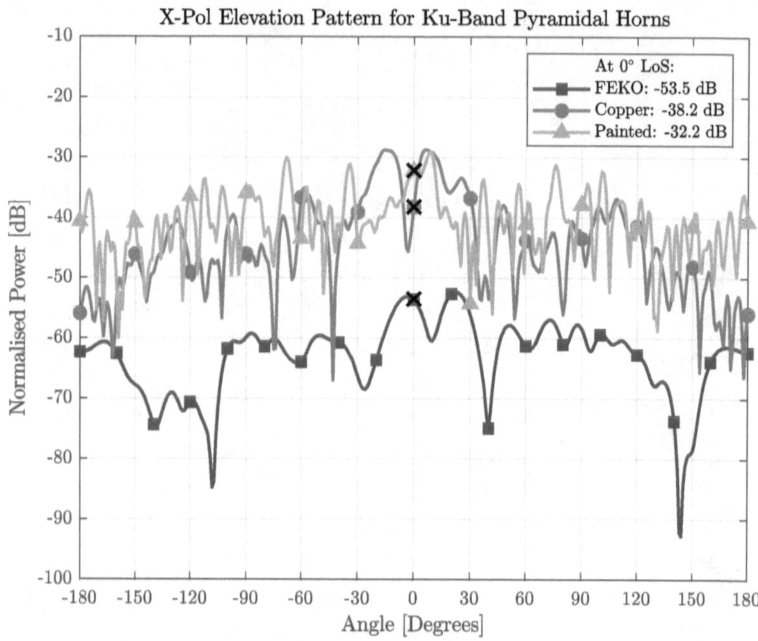

Figure 5-12: Elevation x-pol pattern of Ku-band pyramidal horns for FEKO simulation, copper horn, and painted horn.

The power level at 0° LoS for both the azimuth and elevation pattern of the copper-coated horn are −38.2 dB. For the painted horn, the 0° LoS power levels of the azimuth and elevation patterns are also the same value of −32.2 dB. The fundamental mode of propagation is TE_{10} where the E-field distribution is only in the vertical direction, and therefore has a low x-pol leakage.

5.2.5 Summary

Table 5-2 presents a summary of all the results for the Ku-band pyramidal horn, namely the FEKO simulation, the 3D printed and copper-coated horn, and the 3D printed and painted horn at 15 GHz.

Table 5-2: Summary of antenna properties for the Ku-band pyramidal horns at 15 GHz.

		S_{11}	Gain	HPBW	SLL	X-Pol
FEKO	Azimuth	−51.2 dB	18.8 dBi	20.8°	29.3 dB	−72.3 dB
	Elevation			18.8°	10.7 dB	−53.5 dB
Copper Horn	Azimuth	−23.3 dB	17.7 dBi	21.3°	26.8 dB	−38.2 dB
	Elevation			19.2°	10.5 dB	−38.2 dB
Painted Horn	Azimuth	−24.2 dB	17.5 dBi	21.0°	27.3 dB	−32.2 dB
	Elevation			18.4°	9.9 dB	−32.2 dB

The measured reflection coefficients of the fabricated Ku-band pyramidal horns match well with each other, and the achieved gain compares well with simulation. The azimuth and elevation HPBWs correspond well with the theoretical and simulated values, and the SLLs are at reasonable levels, which indicates the shape of the horn has been constructed well. The measured x-pol values at 0° LoS are below −32 dB in both azimuth and elevation planes. Overall, the measured properties of the fabricated horns match closely with the FEKO simulation.

5.3 Ku-Band Conical Horn

This section presents the measured antenna properties of the Ku-band conical horn antennas. The reflection coefficient, gain, and co-pol and x-pol patterns of the FEKO simulated conical horn, the copper-coated conical horn and the spray-painted conical horn are presented.

5.3.1 Reflection Coefficient of Ku-Band Conical Horn

Figure 5-13 shows the reflection coefficient of the Ku-band conical horns for the FEKO simulated horn, the copper horn, and the painted horn from 14 to 16 GHz. The respective reflection coefficient values at 15 GHz are −50.2 dB, −18.3 dB, and −15.2 dB.

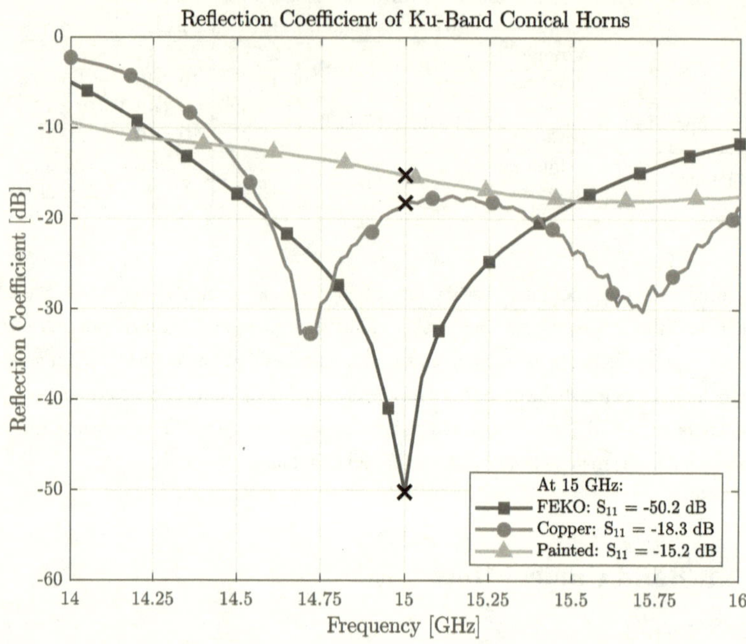

Figure 5-13: Reflection coefficient of Ku-band conical horns for FEKO simulation, copper horn, and painted horn.

The copper horn has a better reflection coefficient performance than the painted horn for frequencies above 14.4 GHz. This is due to the conductivity of the material used on the coating. Copper is a better conductor than the nickel-compound material, therefore it will be more efficient for EM waves to travel through a copper-coated waveguide than a nickel-compound-coated waveguide.

5.3.2 Gain of Ku-Band Conical Horn

Figure 5-14 shows the gain of the FEKO simulated Ku-band conical horn, the copper conical horn, and the painted conical horn from 14 to 16 GHz. The respective gain values at 15 GHz are 17.7 dBi, 16.6 dBi, and 12.0 dBi.

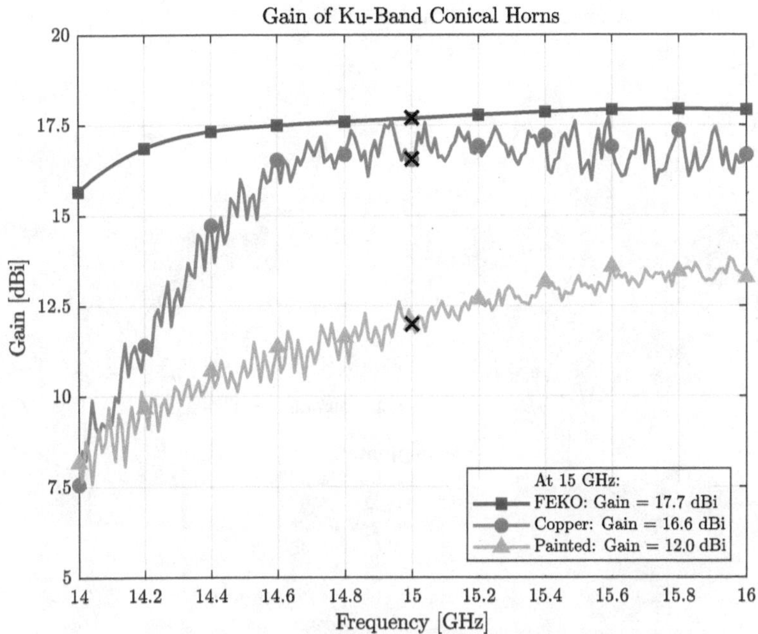

Figure 5-14: Gain of Ku-band conical horns for FEKO simulation, copper horn, and painted horn.

The measured gain of the painted horn at 15 GHz is less than that of the copper horn. This may be a result of the different materials used to metallise the horns, as well as the application of the materials. Copper has a higher conductivity than nickel, and nickel absorbs more energy than copper at higher frequencies. The inner diameter of the circular waveguide is very small, and as a result the application of the spray paint onto the inner surface may have been uneven or too thin, which cause losses in the propagation of the EM wave through the waveguide.

5.3.3   Radiation Pattern of Ku-Band Conical Horn

Figure 5-15 shows the azimuth radiation pattern for the FEKO simulated Ku-band conical horn, the copper conical horn, and the painted conical horn at 15 GHz. The respective azimuth HPBWs are all 24.7°, while the respective azimuth SLLs are 39.7 dB, 23.0 dB and 16.0 dB.

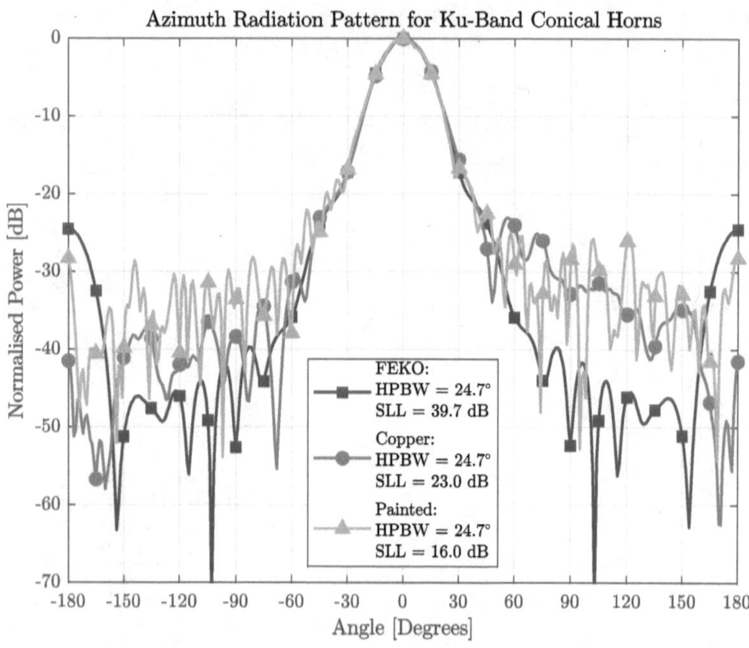

Figure 5-15: Azimuth radiation pattern of Ku-band conical horns for FEKO simulation, copper horn, and painted horn.

Figure 5-16 shows the elevation radiation pattern for the FEKO simulated Ku-band conical horn, the copper conical horn, and the painted conical horn at 15 GHz. The respective elevation HPBWs are 20.5°, 21.1° and 19.0°, while the respective azimuth SLLs are 24.5 dB, 22.1 dB and 10.5 dB.

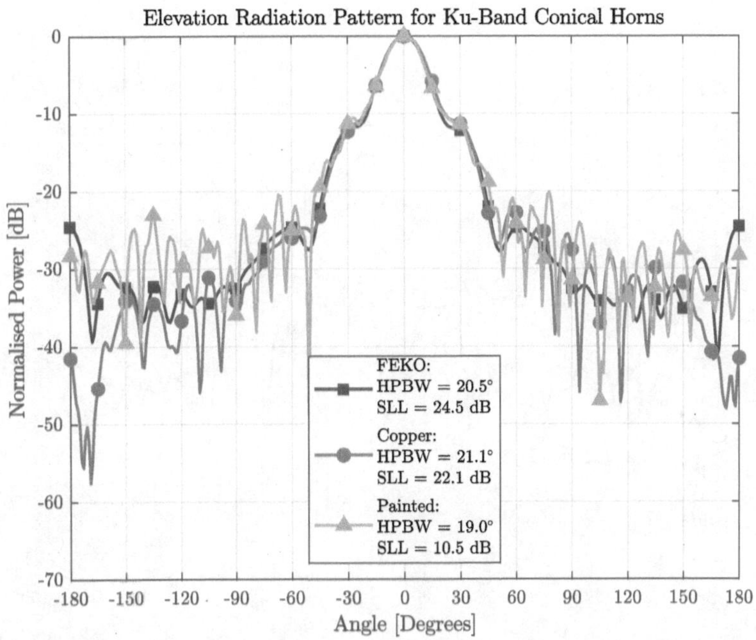

Figure 5-16: Elevation radiation pattern of Ku-band conical horns for FEKO simulation, copper horn, and painted horn.

The elevation HPBWs match very closely to the FEKO simulation, and the SLLs are at acceptable levels.

5.3.4 Cross-Polarisation Pattern of Ku-Band Conical Horn

The x-pol patterns of the Ku-band conical horn are presented. Figure 5-17 shows the azimuth x-pol pattern for the FEKO simulated Ku-band conical horn, the copper conical horn, and the painted conical horn at 15 GHz.

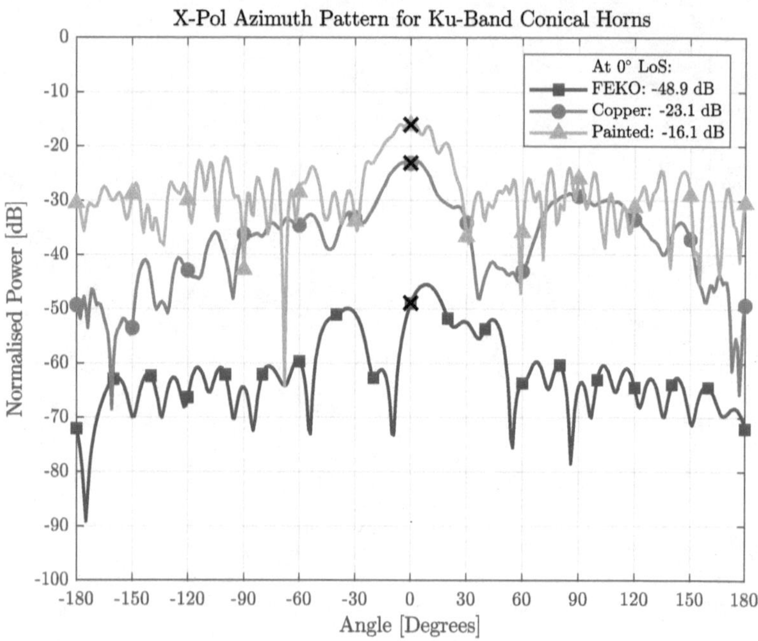

Figure 5-17: Azimuth x-pol pattern of Ku-band conical horns for FEKO simulation, copper horn, and painted horn.

In Figure 5-17 the simulated measurement of the x-pol has a better response than that of the fabricated horns. At 0° LoS, the copper-coated conical horn has a power level of −23.1 dB and the spray-painted horn has a power level of −16.1 dB at 15 GHz. The fabricated horns have a higher x-pol because the simulated measurements are the ideal representations of the antenna properties.

Figure 5-18 shows the elevation x-pol pattern for the FEKO simulated Ku-band conical horn, the copper conical horn, and the painted conical horn at 15 GHz. At 0° LoS, the copper-coated conical horn has a power level of −23.1 dB and the spray-painted horn has a power level of −16.1 dB at 15 GHz.

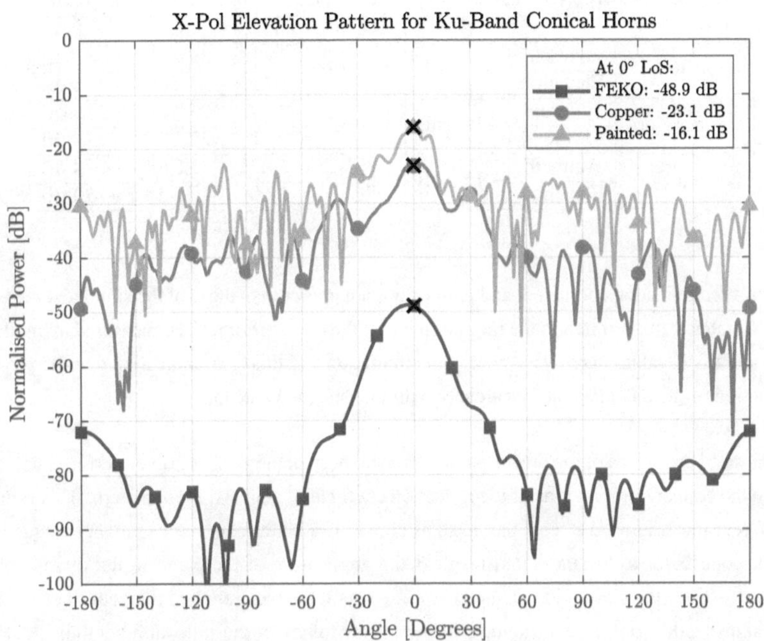

Figure 5-18: Elevation x-pol pattern of Ku-band conical horns for FEKO simulation, copper horn, and painted horn.

The fundamental mode of propagation is TE_{11} where the E-field distribution curves towards the edges, having vertical and minor horizontal E-field components which contributes to the x-pol leakage.

5.3.5 Summary

Table 5-3 presents a summary of all the results for the Ku-band conical horn, namely the FEKO simulation, the 3D printed and copper-coated horn, and the 3D printed and spray-painted horn.

Table 5-3: Summary of measured antenna properties for the Ku-band conical horns at 15 GHz.

		S_{11}	Gain	HPBW	SLL	X-Pol
FEKO	Azimuth	−50.2 dB	17.7 dBi	24.7°	39.7 dB	−48.9 dB
	Elevation			20.5°	24.5 dB	−48.9 dB
Copper Horn	Azimuth	−18.3 dB	16.6 dBi	24.7°	23.0 dB	−23.1 dB
	Elevation			21.1°	22.1 dB	−23.1 dB
Painted Horn	Azimuth	−15.2 dB	12.0 dBi	24.7°	16.0 dB	−16.1 dB
	Elevation			19.0°	10.5 dB	−16.1 dB

The measured reflection coefficient and gain values are poorer than those of the simulated measurements, which is demonstrated in the measurement of the conical horns. The measured azimuth and elevation HPBW values agree very well with simulation and this is an indication that the physical construction of the antenna matches closely with the designed antenna.

The measured properties of the nickel-painted conical horn demonstrates poorer performance compared to the copper horn in terms of the reflection coefficient, gain and x-pol pattern. This is due to the different material used to coat the horn, as copper is a better conductor at higher frequencies. Another contributor to the measured result is the application of the paint on the interior of the waveguide. Since the waveguide diameter is quite small, the paint applied may have been uneven or the skin depth too thin, contributing to the power losses in the reflection coefficient. However, both fabricated horns still compare fairly well with the simulated design, which validates the fabrication method used to construct the antennas.

Chapter 6

Conclusions and Recommendations

6.1 Conclusions

The feasibility of constructing a basic antenna by the means of additive manufacturing with metallisation has been evaluated in this project. Relevant theory and literature have been reviewed, and basic pyramidal and conical horn antennas have been designed, fabricated, coated, and their electrical properties measured. With the limitation of available resources, the overall experiments and measurements showed very promising results, and the 3D printing and conductive coatings have been considered successful. Additionally, the entire fabrication process is very low-cost in comparison to the costs of traditional manufacturing.

6.1.1 X-Band Pyramidal Horn

The first approach involved replicating a commercial X-band pyramidal horn antenna which was readily accessible within the department. The geometric dimensions have been measured and used to perform a simulation in FEKO. A 3D model of the horn has been constructed and 3D printed using the Ultimaker 2+ 3D printer. The antenna has been metallised using a nickel based conductive spray paint, NSCP. The antenna properties of the commercial X-band horn and the fabricated X-band horn have been measured using the anechoic chamber facilities at SU. The measured antenna properties match very closely with the FEKO simulation of the commercial horn. More importantly, the measurements of the replicated spray-painted horn aligns very well with the commercial horn, which concludes that the objectives of this research project have been met. Additional designs of pyramidal and conical antennas and an additional coating method (copper) have been simulated, fabricated and measured to further test the research objective.

6.1.2 Ku-Band Pyramidal and Conical Horns

Another two antennas have been designed in the Ku-band, namely the pyramidal horn and the conical horn. 3D models of the horns have been fabricated using the 3D printer, and both the pyramidal horn and the conical horn have been metallised using two methods: copper-plating performed by TraX Interconnect, and coating with NSCP. All four Ku-band antennas have been measured in the SU anechoic chamber, and the measured antenna properties have been compared with the FEKO simulations.

The comparisons show that the fabricated horns achieved a higher reflection coefficient than simulation, but have all been below $-15\,$dB which means only $3.16\,\%$ of the power is being reflected back by the antenna (or $96.84\,\%$ of the power transmitted is being radiated, which also includes losses in the antenna). The gain achieved by the painted horns have been lower than that of the copper horns, which is due to the conductive properties of the materials used, copper has a higher conductivity compared to nickel at the same frequency of $15\,$GHz. The achieved azimuth and elevation HPBWs are very closely matched to the FEKO simulations.

Comparing the x-pol patterns for the Ku-band pyramidal and conical horns, it is seen that the conical horn has more x-pol leakage. This corresponds with the literature because the fundamental mode of propagation for the pyramidal horn is TE_{10} and for the conical horn is TE_{11}. The TE_{11} mode shows that the E-field distribution curves towards the edges, having vertical and minor horizontal E-field components and thus contributes to a larger x-pol leakage. In comparison, the E-field distribution in the TE_{10} are only in one direction (vertical), and therefore has a much lower x-pol leakage.

The FEKO simulated measurements of the antenna properties show better performance compared to the measurements of the fabricated horns. The largest contributor to the superior performance is because FEKO is a simulator that relies on mathematical algorithms which does not accurately reflect a realistic environment. FEKO also constructs perfect antenna shapes, and the optimisation feature can determine the best probe length and position for optimum performance of the reflection coefficient at the desired frequency. However, FEKO simulation still provides an excellent reference in designing antennas and the flexibility of changing variables to achieve the required performance before the fabrication stage.

Both horn antennas provide very directional beams and high gain. The benefit of the pyramidal horn is that it is more suitable if low x-pol is needed, but it is also more difficult to metallise due to sharp edges. The conical horn has a symmetrical plane which is suited for dual-polarisation and circular polarisation, and easier to metallise. However, in contrast to the pyramidal horn, it has a higher x-pol leakage.

6.1.3 FDM-Based 3D Printer: Ultimaker 2+

3D printing of the horns has been achieved using the Ultimaker 2+ FDM-based 3D printer and ABS is the material chosen for printing. Compared to PLA, ABS is more difficult to work with as a 3D printing filament, but ABS has a higher temperature tolerance, and is more widely used as a plastic in metallisation processes. One of the difficulties of printing in ABS on the Ultimaker 2+ is the adhesion of the part to the printing platform. Poor bed adhesion causes the part to curl upwards along the border of the print. To prevent this, a skirt has been added to the first two layers of the print to improve the adhesion of the part to the printing platform. Because ABS needs a higher temperature to melt, it is also crucial to ensure that the temperature of the printing environment remains constant. If the current layer of the print cools down too much, the next layer will not adhere well to the print and there will be cracks between the layers. This has been prevented by enclosing open windows of the printer with cardboard to regulate the bed and print temperature.

6.1.4 Metallisation by Plating and Conductive Painting

The electroless plating stage of the copper-plating process performed by a local printed circuit board manufacturer, TraX Interconnect, has been a new experimental process for the company, and as such the service has been performed free of charge. The resulting copper-coated Ku-band pyramidal and conical horns have been well-coated, except for some holes which have been more prominent along sharp edges, such as the edges along the flare length of the pyramidal horn. The holes and other discontinuities in the copper-coating have been remedied by applying copper tape over the holes.

The nickel-based conductive spray paint used on the X-band and Ku-band horns has been a relatively simple method of metallisation compared to plating. The process is faster and simpler to perform, and the pyramidal horn antennas metallised using this method achieved slightly poorer reflection coefficient and gain measurements. The conical horn achieved lower than expected performance compared to the pyramidal horns, and this is due to the possible insufficient paint coverage on the inner dimension of the horn. Overall the horn antennas still compare fairly well with simulations.

6.1.5 Cost Evaluation

The total cost of making the X-band pyramidal horn and the Ku-band pyramidal and conical horns amounted to ZAR 2366.96. It is estimated that each horn costs approximately ZAR 475, which includes the cost of fabrication and additional parts needed for measuring the antenna properties. This is significantly cheaper than fabrication by means of CNC machining or even metal 3D printing. Metal Heart, a South African additive manufacturing company, uses the SLM process to 3D print metal parts and it would cost approximately ZAR 5000 or more for them to fabricate the designed Ku-band pyramidal horn. The method of fabrication evaluated in this project is a fast and inexpensive process and would be suited for the design and prototyping stages of components.